DYNAMICAL SYMMETRY OF THE KEPLER-COULOMB PROBLEM IN CLASSICAL AND QUANTUM MECHANICS: NON-RELATIVISTIC AND RELATIVISTIC

DYNAMICAL SYMMETRY OF THE KEPLER-COULOMB PROBLEM IN CLASSICAL AND QUANTUM MECHANICS: NON-RELATIVISTIC AND RELATIVISTIC

TAMAR T. KHACHIDZE
AND
ANZOR A. KHELASHVILI

Nova Science Publishers, Inc.
New York

For permission to use material from this book please contact us:
Telephone 631-231-7269; Fax 631-231-8175
Web Site: http://www.novapublishers.com

LIBRARY OF CONGRESS CATALOGING-IN-PUBLICATION DATA

Khachidze, Tamar T.

Dynamical symmetry of the Kepler-Coulomb problem in classical and quantum mechanics : non-relativistic and relativistic / Tamar T. Khachidze and Anzor A. Khelashvili (authors).

p. cm.

ISBN 978-1-60456-499-0 (hardcover)

1. Symmetry (Physics) 2. Mechanics. 3. Quantum theory. I. Khelashvili, A. A. II. Title.

QC174.17.S9K4 2008

539.7'25--dc22 *2008005561*

Published by Nova Science Publishers, Inc. ✦ New York

CONTENTS

ABOUT THE AUTHORS

Tamar Teymuraz Khachidze (b.1981) graduated from Ivane Javakhishvili Tbilsi State University (TSU) in 2005.

She earned her Master's and PhD degrees in Physics at TSU under the supervision of Professor A.Khelashvili. Now she is working at the High Energy Physics Institute of TSU.

Professor Anzor Alexander Khelashvili (b.1938) is a Doctor of Sciences in Theoretical and Mathematical Physics, graduated from TSU. PhD degree earned at Joint Institute for Nuclear Research (Dubna, Moscow Region).

Professor A. Khelashvili is an author of more than 120 scientific articles, which are devoted to the principle topics of symmetries, many-body problems and composed systems in the framework of QCD, standard model, theory of nonlinear phenomena and topological models in quantum field theory. He is an author of several monographies and teaching books in general and theoretical physics.

PREFACE

This monograph summarizes the major developments that have taken place in the problem of hidden or dynamical symmetries in classical and quantum mechanics (both non-relativistic and relativistic). The findings of Joseph Lois Francois Bertrand (1822-1900) about central fields for which all bounded orbits are closed, was comprehended in the twentieth century as a manifestation of additional symmetry, which is a characteristic feature of two potentials only – namely, Coulomb and isotropic harmonic oscillator. The corresponding conserved quantity, called Laplace-Runge-Lenz (LRL) vector, governs a wide variety of physical systems ranging from planetary motion to atomic spectra.

In our book, the main attention is given to the Kepler-Coulomb problem, while the oscillator is also considered. This is caused by the fact that the Kepler-Coulomb problem has more practical applications: closeness of orbits and degeneracy of motion in classical (Newton's) mechanics, "accidental" degeneracy of spectrum in the hydrogen atom, etc.

In classical relativistic mechanics the symmetry connected to the LRL vector is spoiled, but some traces of it still remain, because in selected rotating reference frame orbits keep closed.

Naturally, there remain some traces also in relativistic quantum mechanics, especially in the Dirac equation. For this case the additional symmetry (in terms of conserved Johnson-Lippmann operator) also takes place. In parallel, the new kind of symmetry—namely, supersymmetry—appears.

It is very interesting that, under the requirement of N=2 supersymmetry of the Dirac Hamiltonian, general central interactions unambiguously follow the Coulomb potential only.

The book is organized as follows:

The first four chapters contain extended reviews of the main results about dynamical symmetries and corresponding constants of motion in non-relativistic and relativistic classical mechanics and non-relativistic quantum mechanics.

The authors' original results are discussed in the remaining chapters. They are devoted to dynamical symmetries, especially supersymmetry in relativistic quantum mechanics, i.e., in the Dirac equation. We have tried to write this book as self-contained as possible.

In spite of the fact that there are many excellent review papers on this subject, this book is a first attempt to collect main results in this field. Therefore, naturally it may not be free of some defects. The authors will be happy to receive any constructive comments and remarks hear readers.

It is remarkable that while there is a very long history of the problems under consideration, new investigations in this direction continue to the present day. There are a lot of papers on monopoles and many other exotic problems. They, unfortunately, are not considered in our book and, therefore, the authors apologize to other researchers if their articles did not fall within our attention.

We would like to take this opportunity to thank our colleagues at the department of Theoretical Physics, Ivane Javakhishvili Tbilisi State University.

Any suggestions for improvement of this book would be greatly appreciated.

We dedicate this book to our students.

Tamar Khachidze
and Anzor Khelashvili

January 2008

INTRODUCTION

"...although the symmetries are hidden from us,
we can sense that they are latent in nature,
governing everything about us. That's the most
exciting idea I know: that nature is much
simpler than it looks"

S. Weinberg

THE GENERAL CONCEPTS OF DYNAMICAL SYMMETRIES

Symmetries play a leading role in modern physics. They determine all regularities of motion, conservation laws and related selection rules. Methods of symmetries are very effective even when the dynamics of interaction is not known well. It has been just symmetries which made possible the unprecedented developments in elementary particle physics, where based on symmetry considerations the unified picture of all interactions was formulated and the further development of theory proceeds to the same ways.

At energies available nowadays the most of symmetries are violated and one of the major tasks of theoretical physics is to clear up what symmetries stay behind this broken picture.

In this respect almost all symmetries are hidden from us. But in physics the notion of "hidden symmetry" has its own significance, which will be defined more precisely below.

It is well known, that symmetries, conservation laws and degeneracy of motion are closely related to each others. Interplay between symmetries and corresponding conservation laws (integrals of motion) is contained in the well-known Noether's theorem [1], according to which if for some coordinate transformations the system's action remains unchanged, than to each symmetry transformation corresponds some integral of motion)[1]. This theorem is valid in classical as well as in quantum mechanics.

On the other hand in Hamiltonian dynamics any integral of motion at the same time is a generator of infinitesimal transformations of underlain symmetry. The Poisson brackets in classical mechanics (respectively, commutators in quantum mechanics) of conserved

---

[1] Naturally, the term "coordinates" means not only space–time variables, but any other, so-called, "internal symmetry" variables as well.

quantities with the Hamiltonian are vanishing. Sets of symmetry generators, as a rule, form some algebras relative to Poisson brackets (commutators).

The character of degeneracy of motion is directly related to symmetries. For example, it is well known, that if a system has a symmetry with respect to spatial rotations, its motion is indifferent to the directions of corresponding conserved quantity – angular momentum vector $\vec{L}$.

Degeneracy of motion by itself may be of various origin. Besides of abovementioned kind of degeneracy, related to coordinate transformations, there are other kinds of degeneracies. Unlike mentioned kind of degeneracy their characteristic feature is that the solution of equations of motion may be carried out by different ways, or more precisely, the problem may be solved in various different coordinate systems.

The Kepler problem, owing to its significant role in the major developments of physics in the last three centuries is probably the most well-known scientific problem in western civilization. It can be solved as in polar (spherical) coordinates, as in so-called, parabolic coordinates [2]. It is clear from general point of view that this degeneracy has to be connected to some symmetry. These new kinds of symmetry are considerably different from the "usual" symmetries, related to geometric transformations and which by their nature are geometrical.

Let us adduce the following quotations from the excellent review article written by H.V. McIntosh [3], in which the essence of dynamical symmetries and their relation to the problem of degeneracy is revealed remarkably:

> "There has been no difficulty in exploiting ostensible geometric symmetry, which is manifested by a group of linear operators commuting with the Hamiltonian of the system. Schur's lemmas describe the limitations imposed on the Hamiltonian, which are substantially that there be no matrix elements connecting wave functions of different symmetry types, and that all the eigenvalues belonging to irreducible representation of the symmetry group be equal. This last mentioned requirement is, of course, the well-known relationship between symmetry and degeneracy. Every symmetric system will show characteristic degeneracies, whose multiplicity is prescribed by the dimensions of the irreducible representations of its symmetry group. Yet, there is no restriction arising from group theoretical reasoning which prevents there from being a higher multiplicity of degeneracy than that required by Schur's lemmas, but any degeneracy so arising is commonly called "accidental" degeneracy due to a prescription as to its unlikelihood...
>
> ...In practice a highly intriguing situation has been noticed. For a great number of the highly idealized and supposedly fundamental systems, there has always been far more degeneracy present than was required by the geometrical symmetry group and Schur's lemmas. For the most part of ostensible symmetry has been the spherical symmetry of central forces in ordinary three-dimensional space, which has been known to require nothing more than a degeneracy in the z-component (direction) of the angular momentum of the wave functions of these systems. The following most typical and extensively treated systems, the hydrogen atom and the harmonic oscillator exhibited degeneracy for various combinations of quantum numbers, resulting in degeneracy which is truly accidental in the context of spherical symmetry...
>
> ...There was always been a filling that accidental degeneracy might not be so much of an accidental after all, in the sense that there might actually have been a larger group which would incorporate several different degenerate representations of the overt symmetry group in a single one of its own irreducible representations.

...Although linear operators, their symmetry groups, and degeneracies are the proper province of quantum mechanics, in any discussion of constants of motion and canonical transformations, classical mechanics will quickly enter the scene, if for no other reason that the fact that most of the concepts and results which are valid in classical mechanics have a fairly immediate transcription into quantum mechanics. Here it must be remembered that historically symmetry has played an important role in classical mechanics as well, although mostly the use of continuous groups of transformations, rather than their matrix representations. Again the basic concept is that of a canonical transformation – a transformation of the phase space variables which leaves unchanged the Hamiltonian form of equations of motion. Among the totality of such transformations there are those which leave the Hamiltonian itself unchanged. The preservation of the Hamiltonian is manifested in two ways: on the one hand, its functional form remains intact after the substitution of the new variables; on the other, when the canonical transformation is an infinitesimal transformation it may be thought of as the generator of a one-parameter Lie group of transformations. This parameter defines an orbit, consisting of the displacement of the point for varying values of the parameter. For example, the z -component of the angular momentum generates, about z -axis, rotations whose orbits are circles orthogonal to one centered on the z -axis; the square of the angular momentum generates, about the angular momentum vectors, rotations which rotate each point in its own plane of motion, which is orthogonal to its angular momentum vector.

...Such is our understanding of symmetry generated by a constant of the motion, the analytic test for such a constant, when it contains no functional dependence on the time, is that its Poisson bracket with the Hamiltonian be zero. However, this is a mutual relationship and inasmuch as the same Poisson bracket describes the temporal variation of the generator, one concludes that this variation must also be zero. Hence, the generators of a symmetry group for the Hamiltonian are also constants of motion. Rather similar considerations apply to quantum mechanical operators.

...The requirement then is for a "hidden" (or dynamical) symmetry: a symmetry not necessarily of a geometric nature, but which together with the geometric symmetries already known would yield a group large enough that its irreducible representations would account for exactly all the observed degeneracies of the system. Classical Hamiltonian mechanics actually contains a reasonable source of hidden symmetries because it deals with a phase space of double the dimension of the configuration space in which the geometric symmetries are evident. In other words, it might be entirely possible that there are additional symmetries of the phase space as a whole which would comprise the desired group".

Thus, as was mentioned above many times, these special kinds of symmetry take place only for special kinds of interactions (dynamics). Therefore they are named as *dynamical symmetries*, underlined the fact that they follow from the special form of equations of motion – only for special kind of forces.

In physics (both classical and quantum) the special role belongs to the two kinds of forces: Coulomb and elastic (isotropic oscillator). Both of them play special role in the nature. Coulomb (gravitational) forces govern planetary motion (Kepler problem), motion of charges in atoms. As regards of elastic forces (Hooke's law) the area of their action is more restrictive - they play essential role only at small distances, they determine the oscillations of atoms and molecules near the equilibrium position. They have a wide application in condensed matter physics, nuclear physics. They may have some relation to the confinement problem for quarks inside hadrons.

In this monograph main attention with be sharpen to the study of dynamical symmetry for Kepler-Coulomb problem, both in classical and quantum mechanics (non-relativistic and relativistic).

REFERENCES

See any textbook on Quantum Field Theory, for example, Bogolyubov N.N. and Shirkov D.V. *Introduction of the Theory of Quantized Fields.* New York, Wiley, 1980.

Landau L.D. and Lifschitz E.M. *Mechanics,* 3rd edition (1982). Butterworth- Heinemann.

McIntosh H.V. *"Symmetry and Degeneracy"*, in *Group Theory and its Applications* (ed. E.M.Loebl), Vol.II, pp.75-144. Acad.Press, New York, 1971.

Chapter I

HIDDEN (DYNAMICAL) SYMMETRIES IN CLASSICAL MECHANICS

"...Both Science and Religion rejoices
in God's symmetry...the endless
contest of light and dark."
Dan Brown

As was pointed out in the introduction symmetries in physics, conservation laws and the degeneracy of motion are interrelated strongly. This is evident in consideration of infinitesimal canonical transformations in classical mechanics. Below we will try to elucidate these relations.

I.1. CONSTANTS OF MOTION AS GENERATORS OF INFINITESIMAL TRANSFORMATIONS

Canonical transformations in classical mechanics are defined as such transformations which preserve the form of Hamiltonian's canonical equations of motion. It means the following : if we perform transition from old canonical variables (q_i, p_i) to the new ones

$$Q_i = Q_i(q, p; t)$$
$$P_i = P_i(q, p; t) \quad (i = 1, 2, ..., n) \tag{I.1}$$

where n is the number of degrees of freedom, the Hamilton equations must have the canonical form again:

$$\dot{Q}_i = \frac{\partial K}{\partial P_i}, \quad \dot{P}_i = -\frac{\partial K}{\partial Q_i} \tag{I.2}$$

here K is a new Hamiltonian.

According to canonical transformations, there is settled definite relations between the old and new variables. For instance, if the canonical transformation is performed by the generating function $F_2(q,P;t)$, then relevant equations are

$$P_i = \frac{\partial F_2}{\partial Q_i}, \; Q_i = \frac{\partial F_2}{\partial P_i} \text{ and } K = H + \frac{\partial F_2}{\partial t} \tag{I.3}$$

for identity transformation

$$F_2 \equiv F_2^0 = \sum_{i=1}^{n} q_i P_i$$

One of the important properties of canonical transformations is that they retain the Poisson brackets. The last ones for 2 independent functions f and g of the canonical variables, are defined as

$$\{f,g\} = \sum_{i-1}^{n} \left(\frac{\partial f}{\partial q_i}\frac{\partial g}{\partial p_i} - \frac{\partial f}{\partial p_i}\frac{\partial g}{\partial q_i} \right) \tag{I.4}$$

The time derivative of arbitrary f function is expressed by Poisson bracket in the following way:

$$\frac{df}{dt} = \frac{\partial f}{\partial t} + \{f,H\} \tag{I.5}$$

when dynamical quantity f does not depend explicitly on time,

$$\frac{df}{dt} = \{f,H\}$$

The infinitesimal transformations are written as follows

$$Q_i = q_i + \delta q_i, \; P_i = p_i + \delta p_i \tag{I.6}$$

Corresponding generating function F_2 must be deviated from identity transformation by infinitesimal amount:

$$F_2 = F_2^0 + \varepsilon G(q,P)$$

Therefore in the first order of variation one can derive the following relations

$$\delta q_i = \varepsilon \frac{\partial G(q,p)}{\partial p_i}, \ \delta p_i = -\varepsilon \frac{\partial G(q,p)}{\partial q_i} \tag{I.7}$$

while for variation of an arbitrary $u(p,q)$ function it is found that

$$\delta u = \varepsilon\{u, G\}$$

In particular case, when u is Hamiltonian, we have

$$\delta H = \varepsilon\{H, G\}$$

when G be a first integral of motion not depended on time explicitly its full time derivative, which is zero, coincides to the Poisson bracket with Hamiltonian, i.e.

$$\{H, G\} = 0 \tag{I.8}$$

It means that the canonical transformation generated by such function G does not change the Hamiltonian. In other words, *every first integral of equation of motion is a generating function of those infinitesimal canonical transformations, which do not change the Hamiltonian.*

Because of this property this kind of transformations transfers every solution with a given energy to other solution with the same energy, i.e. the degeneracy and constants of motion are related to each others: existence of the ones means existents of the others. We will be convinced in following that the constants of motion related to the hidden (dynamical) symmetry are generators of algebra of some higher symmetry group of transformations. In classical mechanics they generate algebras of higher symmetry groups under Poisson brackets. It appears that the additional symmetry beyond the ordinary 3-dimensional rotations in case of central forces takes place only for two potentials, namely, for Coulomb (Newton's) and isotropic oscillator.

At the same time by explicit solution of Newton's equations of motion it was established that for these two potentials the periodic motion on closed orbits are derived. They were derived by Newton himself in 17th century. {I.Newton. Philosophiae Naturalis Principia Mathematica. 1687}, but in 19th century, in 1873 year J.Bertrand prove that only for these two central potentials are derived the periodic motion on closed orbits. Bertrand's theorem has big importance in classical mechanics. The proof of this statement is done in many textbooks and papers.

But for completeness we include here Bertrand's paper, which was written in French, but for its significance English translation appears recently. {arXiv: 0704.2396v1 [physical, class-ph] 18 April 2007}

The translation: ANALYTICAL MECHANICS. - *A Theorem relative to the motion of a point pulled towards a fixed centre; by Mr.* J.Bertrand.

The planetary orbits are closed curves; this is the main cause of the stability of our system, and this important circumstance stems from the law of attraction which, whatever the initial circumstances, makes each celestial body which is not expelled from our system follow the circumstance of an ellipse. Until now it was not observed that the Newtonian law of attraction is the only one that fulfils this condition.

Among the laws of attraction that assume to be null the action at an infinite distance, that of nature is the only one for which a mobile body *arbitrarily* launched with a speed less that a certain limit and pulled towards a fixed centre, describes necessarily a closed curve about this centre. All laws of attraction *allow* closed orbits, but the law of nature is the only one that *imposes* them.

We prove this theorem in he following way: let $\varphi(r)$ be the attraction exerted at a distance r on the molecule (*by molecule Bernard is certainly referring to a particle)* and directed to the centre of the attraction that we will take as the origin of the coordinates. Denoting by r e θ the two polar coordinates of the mobile body; we have by virtue of a well known formula,

$$\varphi(r) = \frac{\kappa^2}{r^2}\left(\frac{l}{r} + \frac{d^2}{d\theta^2}\frac{1}{r}\right)$$

and, setting $\frac{1}{r} = z$.

$$r^2\varphi(r) = \psi(z) \tag{1}$$

$$\frac{d^2 z}{d\theta^2} + z - \frac{1}{\kappa^2}\psi(z) = 0$$

Let us multiply both members by $2dz$ and let us integrate setting

$$2\int\psi(z)dz = w(z), \tag{2}$$

we will have

$$\left(\frac{dz}{d\theta}\right)^2 + z^2 - \frac{1}{\kappa^2}w(z) - h = 0$$

h being a constant.

From this one deduce that

$$d\theta = \pm\frac{dz}{\sqrt{h + \frac{1}{\kappa^2}w(z) - z^2}}$$

If the curve represented by the equation that ties z to θ is closed, the value of z will have maxima and minima for which $dz/d\theta$ will be null and the corresponding vector radii, normal to the trajectory, will necessarily be axes of symmetry for it. Now when a curve admits two axes of symmetry, necessary and sufficient condition for it to be closed is that its angle be commensurable with π. Therefore, if α and β represent a minimum of z and the maximum that follows it, the condition required is expressed by the equation

$$m\pi = \int_{\alpha}^{\beta} \frac{dz}{\sqrt{h + \frac{1}{\kappa^2} w(z) - z^2}} \tag{3}$$

where m denotes a commensurable number. This equation must hold whatever h and κ might be and consequently the limits α and β that depend on them.

One has

$$h + \frac{1}{\kappa^2} w(\alpha) - \alpha^2 = 0$$

$$h + \frac{1}{\kappa^2} w(\beta) - \beta^2 = 0$$

Consequently

$$\frac{1}{\kappa^2} = \frac{\beta^2 - \alpha^2}{w(\beta) - w(\alpha)}$$

$$h = \frac{\alpha^2 w(\beta) - \beta^2 w(\alpha)}{w(\beta) - w(\alpha)}$$

and equation (3) becomes

$$m\pi = \int_{\alpha}^{\beta} \frac{dz\sqrt{w(\beta) - w(\alpha)}}{\sqrt{\alpha^2 w(\beta) - \beta^2 w(\alpha) + (\beta^2 - \alpha^2) w(z) - z^3 [w(\beta) - w(\alpha)]}}. \tag{4}$$

The function $w(z)$ must be such that this equation holds for all values of α e β. Moreover, the commensurable number m must be constant, for if it were to vary from one orbit to another one, an infinitely small variation of the initial conditions would bring forth a finite variation of the number and the disposition of the axes of symmetry of the trajectory.

Assume that α and β differ infinitesimally; let

$$\beta = \alpha + u$$

z staying included between α and β, we can set

$$z = \alpha + \gamma,$$

and γ will be, just as u ,infinitely small. Neglecting the infinitely small of second order we will have

$$\sqrt{w(\beta) - w(\alpha)} = \sqrt{uw'(\alpha)}$$

In the expression under the radical sign in the denominator of the integral (4) the infinitely small of first order reduce to zero, and the same happens with those of second; it is those of third that are necessary to keep, and neglecting the infinitely small of fourth order one has

$$\alpha^2 w(\beta) - \beta^2 w(\alpha) + (\beta^2 - \alpha^2) w(z) - z^3 [w(\beta) - w(\alpha)] =$$
$$= [w'(\alpha) - \alpha w''(\alpha)](u^2 \gamma - u\gamma^2)$$

Equation (4) becomes

$$m\pi = \int_0^u \frac{d\gamma \sqrt{w'(\alpha)}}{\sqrt{w'(\alpha) - \alpha w''(\alpha)} \sqrt{u\gamma - \gamma^2}}$$

that is, performing the integration and suppressing common factors

$$m = \sqrt{\frac{w'(\alpha)}{w'(\alpha) - \alpha w''(\alpha)}},$$

or

$$(1 - m^2) w'(\alpha) + m^2 \alpha w''(\alpha) = 0$$

From this one deduce that

$$w'(\alpha) = \frac{A}{\alpha^{1/m^2 - 1}}$$

$$w(\alpha) = A\frac{\alpha^{2-1/m^2}}{2-\frac{1}{m^2}} + B$$

A and B denoting constants.

From the assumed relations between the functions w, ψ e φ it follows that

$$\psi(z) = \frac{A}{2z^{1/m^2-1}}$$
$$\varphi(r) = \frac{A}{2} r^{1/m^2-3}$$

Such is only possible law of attraction, m here denoting any commensurable number; but from this it does not follow that it fulfills all the conditions of the proposition for any m. In fact, one must have for values of α and β,

$$m\pi = \int_\alpha^\beta \frac{dz\sqrt{\beta^{-\frac{1}{m^2}+2} - \alpha^{-\frac{1}{m^2}+2}}}{\frac{\alpha^2}{\beta^{\frac{1}{m^2}-2}} - \frac{\beta^2}{\alpha^{\frac{1}{m^2}-2}} + (\beta^2-\alpha^2)\frac{1}{z^{\frac{1}{m^2}-2}} - z^2\left(\beta^{-\frac{1}{m^2}+2} - \alpha^{-\frac{1}{m^2}+2}\right)} \tag{6}$$

Let us assume initially $1/m^2 - 2$ negative; let us set $\alpha = 0$, $\beta = 1$, the equation becomes

$$m\pi = \int_0^1 \frac{dz}{\sqrt{\frac{1}{z^{\frac{1}{m^2}-2}} - z^2}} = \int_0^1 \frac{z^{\frac{1}{2m^2}-1}dz}{1 - z^{\frac{1}{m^2}}}$$

and the equation (6) yields

$$m\pi = m^2\pi$$
$$m = 1$$

The corresponding law of attraction is

$$\varphi(r) = \frac{A}{r^2}$$

If one assumes $\frac{1}{m^2} - 2$ positive, equation (6) for $\alpha = 1, \quad \beta = 0$,

$$m\pi = \int_0^1 \frac{dz}{\sqrt{1 - z^2}} = \frac{\pi}{2}$$

From this it follows that $m = 1/2$ and the corresponding law of attraction is

$$\varphi(r) = Ar$$

Only two laws therefore fulfill the required conditions that of nature, by which the closed orbit has only one symmetry axis passing through the centre of action, and the attraction proportional to the distance, by which there are two.

Our illustrious correspondent Mr. Chebyshev, to whom I have communicated the precedent demonstration, judiciously made me remark that the theorem, useless nowadays for the so perfect theory of the planets, may be invoked in a useful way in order to extend to the double stars the Newtonian laws of attraction.

Remark

Bertrand's original paper could be found at the following address: http://gallica.buf.fr/ark:/12148/bpt63034n.

We note that the various proofs of this theorem are considered both in textbooks [1,2] and articles [3,4].

Thus, Bertrand's conclusion is that the only power-law central potentials for which the bounded trajectories are closed are $1/r$ (gravitational and Coulomb) and r^2 (isotropic 3-dimentional harmonic oscillator). In other cases of attractive potentials the circular motion is singular. If the circular orbit is slightly distorted, periodicity immediately disappears.

The additional degeneracy in the Kepler problem was, of course, mentioned earlier. For a particle moving in Coulomb potential the total energy is depended only on the value a of large semi-axis, namely, $E = -K/2a$, but not on orbit's eccentricity. By the same reason in quantum mechanics hydrogen atom's energy levels do not depend on orbital quantum number l, while the last one is explicitly present in the radial Schroedinger equation.

Hence, the Bertrand's theorem dictates us that the degeneracy in these potential fields must be provided by the existence of additional integrals of motion.

Indeed, it turned out that in classical mechanics closeness of orbits and strict periodicity of motion on them is provided by additional vector constant, which in scientific literature is known as the Laplace – Runge – Lenz (LRL) vector. This vector is named by names of three authors. It is caused by historical justice. The interested reader may be recommended to the several papers of H. Goldstein[5,6].

It is remarkable, that the derivation of LRL vector is based on performing some manipulations on the equation of motion, but not on some general principles. The reason, probably, is that the symmetry group, generated by this vector, has no ordinary geometrical meaning and corresponds really to "hidden symmetry".

Below the one of the variance of derivation of LRL vector will be done.

I.2. Derivation of LRL Vector

Let us write down the Newton's second law for particle moving in some central field

$$\frac{d\vec{p}}{dt} = f(r)\frac{\vec{r}}{r}, \quad (I.9)$$

where $f(r)$ is the radial value of force:

$$f(r) = -\frac{\partial V(r)}{\partial r} \qquad (I.10)$$

and $\vec{p} = m\vec{\upsilon}$.

Multiplying equation of motion by orbital momentum vector $\vec{L}$, which is constant vector on any central symmetric field and perform suitable transformations; Taking into account, that $\vec{L} = [\vec{r} \times \vec{p}]$ and $\vec{p} = m\dot{\vec{r}}$, we derive:

$$\dot{\vec{p}} \times \vec{L} = \frac{mf(r)}{r}\left[\vec{r} \times \left[\vec{r} \times \dot{\vec{r}}\right]\right] =$$

$$= \frac{mf(r)}{r}\left\{\vec{r}\left(\vec{r} \cdot \dot{\vec{r}}\right) - r^2\dot{\vec{r}}\right\}$$

Further simplification of this equation is possible under accounting that

$$\vec{r} \cdot \dot{\vec{r}} = \frac{1}{2}\frac{d}{dr}\left(\vec{r} \cdot \vec{r}\right) = \frac{1}{2}\frac{d}{dt}\left(r^2\right) = r\dot{r}$$

This simply means that the radial component of velocity $\dot{\vec{r}}$ is $\dot{r}$.

Using constancy of $\vec{L}$, it is found that

$$\frac{d}{dt}\left[\vec{p}\times\vec{L}\right]=-mf(r)r^{2}\left(\frac{\dot{\vec{r}}}{r}-\frac{\dot{r}\vec{r}}{r^{2}}\right)$$

or

$$\frac{d}{dt}\left[\vec{p}\times\vec{L}\right]=-mf(r)r^{2}\frac{d}{dt}\left(\frac{\vec{r}}{r}\right)$$

Now without fixing of $f(r)$ further motion is impossible.

It is evident that integration of this equation will be possible only if $r^{2}f(r)=const$, or $f(r)\sim r^{-2}$, which is exactly so in the Kepler problem,

$$f(r)=-\frac{a}{r^{2}} \tag{I.11}$$

Therefore we obtain

$$\frac{d}{dt}\left\{\left[\vec{p}\times\vec{L}\right]-ma\frac{\vec{r}}{r}\right\}=0$$

or we find a constant vector

$$\vec{A}\equiv\left[\vec{p}\times\vec{L}\right]-ma\hat{\vec{r}}=const., \tag{I.12}$$

where $\hat{\vec{r}}=\frac{\vec{r}}{r}$ is an unit vector along $\vec{r}$.

Among physicists this vector is often called as Runge – Lenz vector, sometimes as Runge –Lenz – Pauli vector, while priority belongs to Laplace.

Laplace elaborated a general method for construction of integrals of motion, which is based on consideration of the most general monomials of coordinates and their time derivatives. Then the time independence is required. This procedure is explained in detail in textbooks and review articles [7].

It is interesting to note that after Laplace monomials with higher than second degrees were not considered till now, may be because of large uncertainties of method.

I.3. Applications of LRL Vector in Classical Physics

(I) LRL Vector and the Orbit Equation

It is clear from the definition, that vector $\vec{A}$ lies in the orbit plane, therefore it is perpendicular to $\vec{L}$:

$$\vec{L}\cdot\vec{A}=\vec{A}\cdot\vec{L}=0 \tag{I.13}$$

It follows that $\vec{A}$ is fixed vector in the orbit plane. If we denote by θ the angle between $\vec{A}$ and radius–vector $\vec{r}$, and consider the scalar product, we find

$$\vec{r}\cdot\vec{A}=rA\cos\theta=\vec{r}\left[\vec{p}\times\vec{L}\right]-mar$$

But the cyclic permutation gives

$$\vec{r}\cdot\left[\vec{p}\times\vec{L}\right]=\vec{L}\left[\vec{r}\times\vec{p}\right]=\vec{L}\cdot\vec{L}=L^2$$

Therefore $rA\cos\theta=L^2-mar$ or

$$\frac{1}{r}=\frac{ma}{L^2}\left(1+\frac{A}{ma}\cos\theta\right) \tag{I.14}$$

So, the LRL vector gives us the different (algebraic) method for obtaining of orbit equation without solving of equation of motion.

If we compare this result to the orbit equation in a familiar form

$$\frac{1}{r}=p(1+\varepsilon\cos\theta) \tag{I.15}$$

we conclude, that $\vec{A}$ is directed along the radius-vector to perihelion of orbit and its magnitude is

$$A=ma\varepsilon \tag{I.16}$$

where ε is an eccentricity of the orbit.

If we calculate the square of LRL vector and take into account that the total energy is

$$E = \frac{p^2}{2m} - \frac{a}{r},$$

we find the relation

$$\vec{A}^2 = m^2 a^2 + 2mEL^2$$

From which the familiar expression [1] for orbit's eccentricity follows

$$\varepsilon = \frac{A}{ma} = \sqrt{1 + \frac{2EL^2}{ma^2}} \quad \text{(I.17)}$$

In other words, within the orbit plane the spatial orientation and the geometrical form of the Kepler orbit are completely determined by the LRL vector.

The only quantity left undetermined about orbit as a whole is its spatial size, or physical scale. This scale is given by the magnitude L of the angular momentum (through the factor $p \equiv L^2/ma$).

(II). Algebraic Aspects of the Kepler Problem

It is interesting to make clear to which symmetry corresponds the LRL vector. For this aim it is most suitable consideration of Poisson brackets between conserved quantities.

The existence of constant angular momentum $\vec{L}$ is related to the invariance of a system Hamiltonian with respect to spatial $O(3)$ rotations. This property is expressed by the following Poisson bracket:

$$\{L_i, L_j\} = \varepsilon_{ijk} L_k \quad \text{(I.18)}$$

At the same time, a conservation of this vector means

$$\{L_i, H\} = 0$$

where H is a Hamiltonian

$$H = \frac{p^2}{2m} - \frac{a}{r}$$

Let us remark that the vanishing of Poisson bracket of L_i with the Hamiltonian takes place for arbitrary central potential $V(r)$.

It is easy to calculate the next bracket

$$\{L_i, A_j\} = \varepsilon_{ijk} A_k \tag{I.19}$$

This relation exhibits simply that $\vec{A}$ is a vector under rotations generated by $\vec{L}$.

The following step is a calculation of the Poisson bracket between components of LRL vector, which gives

$$\{A_i, A_j\} = \sqrt{-2m|H|}\varepsilon_{ijk} L_k \tag{I.20}$$

It is more convenient to choose another normalization of this vector

$$A_i \to D_i = \frac{1}{\sqrt{-2m|H|}} A_i \tag{I.21}$$

It is evident that for negative energies (bound motion) $\vec{D}$ is a real vector. For its components the Poisson bracket simplifies:

$$\{D_i, D_j\} = \varepsilon_{ijk} L_k \tag{I.22}$$

Moreover

$$\{D_i, H\} = 0,$$

which expresses a constancy of D_i.

Therefore, collecting all brackets together, we see that we have two sets of constant vectors, L_i and D_i, which obey the extended algebra under Poisson brackets:

$$\begin{aligned} \{L_i, L_j\} &= \varepsilon_{ijk} L_k \\ \{L_i, D_j\} &= \varepsilon_{ijk} D_k \\ \{D_i, D_j\} &= \varepsilon_{ijk} L_k \end{aligned} \tag{I.23}$$

These 6 generators do not change the Hamiltonian and compose the closed algebraic relations under Poisson brackets. But there is one additional relation between them, which determines orbit. So, only 5 of them are independent.

In order to clarify the symmetry they generate, let us give to this algebra more ordinary form introducing the following linear combinations:

$$J_i = \frac{L_i + D_i}{2}, \; K_i = \frac{L_i - D_i}{2} \tag{I.24}$$

Then above Poisson brackets reduce to two independent subsets

$$\{J_i, J_j\} = \varepsilon_{ijk} J_k, \qquad \{K_i K_j\} = \varepsilon_{ijk} K_k \tag{I.25}$$

$$\{J_i, K_j\} = 0$$

These relations explicitly express two sets of analytically conserved quantities, each of them describing some 3-dimensimal rotations (first line), which are mutually independent (second line).

Therefore, we are dealing here with a direct product of two algebras $0^+(3) \times 0^+(3)$, which is equivalent to the rotations in Euclidean 4-dimensional space, i.e.

$$0(4) \sim 0^+(3) \times 0^+(3)$$

I.4. Dynamical Symmetry for the Isotropic Harmonic Oscillator

It is natural that the construction of dynamical symmetry probably must be possible for second potential, mentioned in Bertrand's theorem – spatial isotropic harmonic oscillator.

Let's begin with the two-dimensional anisotropic oscillator[2]) and take its Hamiltonian in the form

$$H = H_1 + H_2$$

as a sum of two oscillators. Here every H_i is the Hamiltonian, describing oscillators along separate axis

$$H_i = \frac{p_i^2}{2m} + \frac{m\omega_i^2 x_i^2}{2} \quad (i = 1,2) \tag{I.26}$$

As is well-known the generator of rotations around the axis, perpendicular to (x_1, x_2)-plane is the momentum L_3

$$L_3 = \varepsilon_{3jk} x_j p_k = x_1 p_2 - x_2 p_1$$

[2] Let us mention, that oscillators in various dimensions were also considered in literature [8].

Calculating its Poisson bracket with total Hamiltonian, we obtain

$$\{L_3, H\} = \frac{1}{2m} x_1 x_2 \left(\omega_1^2 - \omega_2^2\right)$$

We see that only for isotropic oscillator $(\omega_1 = \omega_2 \equiv \omega)$ we'll have symmetry relative to these rotations

$$\{L_3, H\} = 0$$

Let us introduce the symmetric second rank tensor

$$A_{ij} = \frac{1}{2m}\left(p_i p_j + m^2 \omega^2 x_i x_j\right) \tag{I.27}$$

It is easy to show that

$$\{A_{ij}, H\} = 0$$

So, the components of this tensor are generators of such transformations, which remain the Hamiltonian invariant. Moreover,

$$A_{ij} L_j = 0 \tag{I.28}$$

i.e. components of this tensor are concentrated in orbit plane.

If we introduce the following generators

$$K_1 = \frac{A_{12}}{\omega},\ K_2 = \frac{A_{22} - A_{11}}{\omega},\ K_3 = \frac{L_3}{2}$$

It is easy to convince that they form a closed algebra of Poisson brackets:

$$\{K_i K_j\} = \varepsilon_{ijk} K_k \ \ (i, j, k = 1,2,3) \tag{I.29}$$

So, they correspond to some rotations, but now in more (than three)-dimensional space. We see that in analogy to the Kepler problem symmetry here is also enlarged.
Now if we introduce 3-dimensional isotropic oscillator

$$H = H_1 + H_2 + H_3,\ \omega_1 = \omega_2 = \omega_3 \equiv \omega$$

It appears that components of $A_{ij}(i, j = 1,2,3)$ tensor form vanishing Poisson brackets with H and express some conserved quantities.

In this model we have

$(*)$ Hamiltonian: $H = \frac{1}{2m}\left(\vec{p}^{\,2} + m^2\omega^2\vec{r}^{\,2}\right)$

$(**)$ Conserved tensor $A_{ij} = \frac{1}{m\omega}\left(p_i p_j + m^2\omega^2 x_i x_j\right)$

The physical significance of the conserved tensor A_{ij} may be understood in terms of its eigenvalues and corresponding eigefunctions. Let us take $m = 1$ for simplicity and consider the classical eigenvalue problem [9]

$$\sum_j A_{ij}\upsilon_j = a\upsilon_i \tag{I.30}$$

From the secular determinant we find that A_{ij} has three eigenvalues:

$$\begin{aligned} a^{(1)} &= \frac{1}{2}\left[E + \left(E^2 - \omega^2 L^2\right)^{1/2}\right] \\ a^{(2)} &= \frac{1}{2}\left[E - \left(E^2 - \omega^2 L^2\right)^{1/2}\right] \\ a^{(3)} &= 0 \end{aligned} \tag{I.31}$$

where $E = H = \sum_i A_{ii}$.

It is obvious that $\vec{\upsilon}^{(3)}$, the eigenvector associated with $a^{(3)}$, is in the direction of $\vec{L}$. To discuss the other two eigenvectors a coordinate system is chosen so that the (x, y)-plane is the plane of the orbit, and x, y axes are chosen to coincide with the axes $\vec{i}, \vec{j}$ of the elliptical orbit. Then the solution of equation of motion is

$$\begin{aligned} \vec{r} &= \vec{i}\,D_1\cos\omega t + \vec{j}\,D_2\sin\omega t \\ \vec{p} &= -\vec{i}\,D_1\omega\sin\omega t + \vec{j}\,D_2\omega\cos\omega t \end{aligned} \tag{I.32}$$

The amplitudes D_1 and D_2 are related to the energy and angular momentum by

$$E = \frac{\omega^2}{2}\left(D_1^2 + D_2^2\right) \; L^2 = \left(\omega D_1 D_2\right)^2 \tag{I.33}$$

If the initial moment of our time is chosen so that $D_1 \geq D_2$, then it is easy verified that

$$a^{(1)} = \frac{1}{2}\omega^2 D_1^2, \; a^{(2)} = \frac{1}{2}\omega^2 D_2^2 \tag{I.34}$$

In this special coordinate system A_{ij} tensor is diagonal

$$A_{ij} = \frac{1}{2}\omega^2 \begin{pmatrix} D_1^2, & 0, & 0 \\ 0, & D_2^2, & 0 \\ 0, & 0, & 0 \end{pmatrix} \tag{I.35}$$

and the orbit equation takes the standard form for the equation of an ellipse

$$\left(\frac{x_1}{D_1}\right)^2 + \left(\frac{x_2}{D_2}\right)^2 = 1 \tag{I.36}$$

Thus we see that $a^{(1)}, a^{(2)}$ are $\frac{1}{2}\omega^2$ times the square of the length of the major (minor) axes of the elliptical orbit, and $\vec{\upsilon}^{(1)}, \vec{\upsilon}^{(2)}$ are in the direction of these axes. Consequently, the symmetric tensor completely describes the orientation of the orbit for three-dimensional isotropic harmonic oscillator.

A constant quantity giving a complete description of the orbit is possible only because of the periodic nature of the motion, i.e. the orbit is reentrant.

It is easy to check that we have the following algebra of Poisson brackets

$$\begin{aligned} \{L_i, A_{jk}\} &= \varepsilon_{ijl} A_{lk} + \varepsilon_{ikl} A_{jl} \\ \{A_{ij}, A_{kl}\} &= \left(\delta_{ik}\varepsilon_{jlm} + \delta_{il}\varepsilon_{jkm} + \delta_{jk}\varepsilon_{ilm} + \delta_{jl}\varepsilon_{ikm}\right) L_m \end{aligned} \tag{I.37}$$

Therefore, 3 angular momentum components L_i and 5 (traceless) components of A-tensor form together closed algebraic system under Poisson brackets and correspond to wider symmetry transformations in 8-dimensional space. It is $SU(3)$ algebra.

Because the components A_{ij} are located in plane, one can construct other conserved tensor as well, say

$$T_{sj} = \varepsilon_{iks} A_{ij} L_k + \varepsilon_{ijk} A_{is} L_k$$

This lies also in the same plane. In general the linear combination of tensors A_{ij} and T_{ij} will be also conserved tensor.

It may be shown in general that in n-dimensions isotropic oscillator $(n \geq 2)$ forms $SU(n)$ algebra of Poisson brackets.

Because for considered examples of these two potentials larger symmetry appears mainly as a result of motion in a plane, one may expect, that analogical symmetries can be possible for all central potentials.

I.5. Possible Generalizations of Dynamical Symmetries

In general vector $\vec{D}$ as well as vector $\vec{A}$ lie in orbit plane and are directed along a large axis (for bound states). It follows that $\vec{L} \times \vec{A}$ is also a constant vector in orbit plane, which is directed along a small axis. Therefore the quantity

$$c_1 \vec{A} + c_2 \vec{L} \times \vec{A} \tag{I.38}$$

is a constant vector also in orbit plane for arbitrary c_1, c_2, which determine its direction.

Because the existence of such constant vectors is provided by a constancy of the orbit plane, which is valid for all central potentials, so one can think that it may be possible to find similar to LRL vector for all central potentials.

This problem was solved by Fradkin in 1967. Below we describe the main points of his work [10]. We'll see how a constant vector in orbit plane can be characterized by canonical variables $\vec{r}$ and $\vec{p}$.

Let $\hat{\vec{k}}$ be a unit vector in orbit plane. It is clear, that it might be decomposed according 3 mutually orthogonal directions, $\hat{\vec{r}}$, $\hat{\vec{L}}$ and $\hat{\vec{r}} \times \hat{\vec{L}}$:

$$\hat{\vec{k}} = \left(\hat{\vec{k}} \cdot \hat{\vec{r}} \right) \hat{\vec{r}} + \left(\hat{\vec{k}} \cdot \hat{\vec{L}} \right) \hat{\vec{L}} + \left(\hat{\vec{k}}, \hat{\vec{r}} \times \hat{\vec{L}} \right) \hat{\vec{r}} \times \hat{\vec{L}} \tag{I.39}$$

By assumption

$$\hat{\vec{k}} \cdot \hat{\vec{L}} = 0$$

Moreover, if one chooses $\hat{\vec{k}}$ in such a direction, from which the azimutal angle θ is counted of, then

$$\hat{\vec{r}} \cdot \hat{\vec{k}} = \cos\theta \,, \ (\hat{\vec{k}} \cdot \hat{\vec{r}} \times \hat{\vec{L}}) = \sin\theta$$

Consequently

$$\hat{\vec{k}} = \hat{\vec{r}} \cos\theta + \hat{\vec{r}} \times \hat{\vec{L}} \sin\theta \tag{I.40}$$

We now want to express $\cos\theta$, $\sin\theta$ by means of $\vec{r}$ and $\vec{p}$.
For this aim let return to equations of motion

$$\frac{d\vec{p}}{dt} = -\frac{\partial V}{\partial r} \hat{\vec{r}}$$

The usual integrals of motion are

$$E = \frac{p^2}{2m}, \qquad \vec{L} = \vec{r} \times \vec{p}$$

According to our choice of θ,

$$L = mr^2 \frac{d\theta}{dt}$$

and

$$p^2 = m^2 \left[\left(\frac{dr}{dt} \right)^2 + r^2 \left(\frac{d\theta}{dt} \right)^2 \right]$$

$$\vec{r} \cdot \vec{p} = mr \frac{dr}{dt}$$

From these relations and equation of motion the general solution follows:

$$\cos\theta = f\left(u, \, L^2, \, E\right) \equiv f, \tag{I.41}$$

where $u = \dfrac{1}{r}$.

In other words, $\cos\theta$ is expressed through $\vec{r}$ and $\vec{p}$, because arguments of f are functions of these variables.

As regards of f function, it is expressed in form of following integral:

$$f = \int^{u} \frac{du'}{u'^2}\left[\left(\frac{2(E-V)}{L}\right)^2 - u'^2\right]^{-1/2}$$

which is a solution of the orbit equation

$$\left(\frac{du}{d\theta}\right)^2 = \frac{2m}{L^2}(E-V) - u^2$$

We have at the same time

$$\sin\theta = \frac{\partial f}{\partial u}\frac{\left(\hat{\vec{r}}\cdot\vec{p}\right)}{L}$$

by taking of which we obtain the Fradkin's vector

$$\hat{\vec{k}} = \left(f - u\frac{\partial f}{\partial u}\right)\hat{\vec{r}} + L^{-1}\left(\frac{\partial f}{\partial u}\right)\vec{p}\times\vec{L}$$

By sufficiently tedious calculating one can to show that

$$\left\{E, \hat{\vec{k}}\right\} = 0$$

which says that $\hat{\vec{k}}$ is indeed constant unit vector. Its vector character is confirmed by following algebraic relation

$$\left\{L_i, \hat{\vec{k}}_j\right\} = \varepsilon_{ijk}\hat{\vec{k}}_k$$

It is essential, that one can multiply this unit vector by an arbitrary function of L^2 and E. We derive still constant vector.

Let now consider a particular case of Coulomb potential

$$V = -au \tag{I.42}$$

Integration of the orbit equation can be performed and we obtain

$$f \equiv \cos\theta = \left[2mEL^2 + (am)^2\right]^{-1/2}\left(L^2u - am\right)$$

Substitution of this expression into $\hat{\vec{k}}$ gives us a familiar LRL vector

$$\vec{D} = (-2mE)^{\frac{1}{2}}\left(\vec{p}\times\vec{L} - am\hat{\vec{r}}\right)$$

Comments

The above generalization of LRL vector has one lack: It has an ambiguity, which gives some non uniqueness in construction of group invariants from it. This happens because the integral so obtained is not connected to the geometry of motion. Only algebraic relations are not sufficient carriers of dynamical information.

Therefore in some next papers [11, 12] there were attempts to avoid this ambiguity with the aid of some additional restrictions, for example, requiring that Fradkin's vector becomes zero when we pass circle orbit, where there is no distinguished direction.

Under this requirement the following vector follows [12]

$$\vec{A} = \left[f(E) - L^2\right]^{\frac{1}{2}}\left\{M\left(r, L^2, E\right)\vec{r} + N\left(r, L^2, E\right)\left[\vec{p}\times\vec{L}\right]\right\} \tag{I.43}$$

where

$$N = L^{-2}\left[\frac{2(E-V)}{L^2} - \frac{1}{r^2}\right]^{-\frac{1}{2}} \sin\theta$$

$$M = \left\{\frac{1}{r}\cos\theta - \frac{1}{r^2}\left[\frac{2(E-V)}{L^2} - \frac{1}{r^2}\right]^{-\frac{1}{2}} \sin\theta\right\}$$

and

$$\theta = \int_{r_1} r'^{-2}\left[\frac{2(E-V)}{L^2} - \frac{1}{r'^2}\right]^{-\frac{1}{2}} dr'$$

Here the integration limit r_1 lies in the classically permissible area

$$2(E-V)r^2 - L^2 > 0 \tag{I.44}$$

Vector $\vec{A}$ lies in the orbit plane, has vanishing Poisson bracket with the Hamiltonian and together with angular momentum vector $\vec{L}$ generates closed algebra of $O(4)$ Lie group. There is arbitrary function $f(E)$ in $\vec{A}$, which underlines the ambiguity of derived constant of motion.

The same happens in the attempts of oscillator's generalization.

Generalized LRL vector proposed is only piecewise conserved or alternatively that it is a multi-valued constant.

Nevertheless many authors remained convinced that the LRL vector proposed is true integral of motion because, as verified by Fradkin, the Poisson bracket of this vector with Hamiltonian is zero.

The object, which can be defined solely in terms of the canonical variables ($\vec{r}$ and $\vec{p}$) of the moving particle and which is independent of time, will be called an invariant of motion.

$\hat{\vec{k}}(t)$ is constant "for almost all time", except specific times, when the particle passes some aphelion (perihelion), at which the direction of $\hat{\vec{k}}$ "jumps" by some finite angle to the direction pointing to the most perihelion (aphelion) at a later time. These lead to the dependence of $\hat{\vec{k}}$ on time. This demonstrates that, in general $\hat{\vec{k}}$ changes in time and is therefore not an integral of the motion.

In such a formulation, the condition for a quantity to be an integral of motion is not only that it shall be constant, but also that it should be a single-valued function of the canonical variables.

The constant vector given by Fradkin abruptly reverses its direction in a complete orbital motion, thus it was named as the piecewise-conserved vector.

Therefore we can conclude that the time dependence of $\hat{\vec{k}}$ established above shows that $\hat{\vec{k}}$ is not an integral of motion for an arbitrary central potential and an arbitrary orbit (i.e. arbitrary E and L) in contrast of the Coulomb potential.

Returning to the related question of dynamical symmetry, mentioned in the introduction, the existence of an integral of motion having generalized LRL character is uniquely related to the existence of closed orbits. While Coulomb and Harmonic oscillator potentials have this property for arbitrary L and E, it has been underlined that a general central potential gives rise to closed orbits only for particular values of L and E. Such closed orbits clearly have very definite geometrical symmetry, which, however, is not to be associated with the particular central field alone.

The question as to whether other rotational invariant time–independent potentials have dynamical algebras for quantum systems has been investigated by Truax (14,1980). He demonstrated that isotropic oscillator, Coulomb and constant are special in the sense that their degeneracy algebras are all larger than $O(3)$. All other central potentials possess $O(3)$ symmetry only.

I.6. Application of the Dynamical Evolution of LRL Vector in General Central Case [12]

The LRL vector provides an extremely useful tool for the numerical calculation and analysis of orbits for any force, central or not. It appears that the most of the relations involving the *LRL* vector in the Kepler-Coulomb problem remain true for an arbitrary central force, and this then immediately leads to the desired set of first-order differential equations of motion.

Equations of Motion for General Central Forces

Let us turn to the problem of determining the orbits for an arbitrary central force: the force still has the traditional form

$$\vec{F}(\vec{r}) = -af(r)\vec{e}_r \tag{I.45}$$

Here and in following $\vec{e}_j$ denotes a unit vector in corresponding direction.But $f(r)$ is now an arbitrary function of the radial distance r. Of course, also in this general case the angular momentum $\vec{L}$ is conserved, and the motion is confined to a plane perpendicular to $\vec{L}$ which we choose as the (xy)-plane.

If we take the ordinary definition of LRL vector

$$\vec{A} = \frac{\vec{r}}{r} - \frac{1}{ma}(\vec{p} \times \vec{L})$$

and ask ourselves which of relations familiar of the Kepler problem:

$$\vec{A} \cdot \vec{L} = 0, \quad \vec{A} = \left(1 - \frac{L^2}{mr^2}\right)\vec{e}_r + \left(\frac{L\dot{r}}{a}\right)\vec{e}_\varphi$$

$$\frac{d\vec{A}}{dt} \equiv \dot{\vec{A}} = \left[1 - r^2 f(r)\right]\frac{L}{mr^2}\vec{e}_\varphi$$

$$\vec{r}\vec{A} = rA\cos(\varphi - \varphi_0) = r - \frac{L^2}{ma}$$

$$r = \frac{L^2}{ma}\frac{1}{1 - 2\cos(\varphi - \varphi_0)} \tag{I.46}$$

remain unchanged in the general case, we immediately find that *all of them remain true.*

While equation for $\dot{\vec{A}}$ does not change, the "only" difference now is that the right-hand side does not vanish because $f(r)$ is not equal to $\frac{1}{r^2}$, i.e. the vector $\vec{A}$ for such a general force is no longer constant of motion. But this equation is still very useful because it provides a first-order differential equation for determination of A. Note that with the help of

$$\dot{\varphi} = \frac{L}{mr^2}$$

we can eliminate the time derivative from the equation for $\dot{\vec{A}}$ in favor of the derivative with respect to the asimuthal angle φ. Written in Cartesian coordinates for $\vec{A} = (A_1, A_2, A_3 = 0)$, this than yields

$$A_1' = -\left[1 - r^2 f(r)\right]\sin\varphi$$
$$A_2' = \left[1 - r^2 f(r)\right]\cos\varphi$$

where the prime denotes the φ derivative. Integration of these differential equation will give $\vec{A}$ as a function of φ.

The usefulness of this becomes clear if we now examine last two equations from (I.46), since their derivation depend only on decomposition of $\vec{A}$ and not on the specific force law.

For the orbit equation this means in particular that *locally* the orbits are always conic sections relative to $\vec{A}$ (which varies in time). Writing the equation in the form

$$r = \frac{L^2}{ma}\frac{1}{1 - \vec{A}\cdot\vec{e}_{\varphi}} \tag{I.47}$$

with

$$\vec{A}\cdot\vec{e}_{\varphi} = A_1(\varphi)\cos\varphi + A_2(\varphi)\sin\varphi$$

This shows that the solution of the differential equations for A_1', A_2' completely determines the orbit equation as a function of φ. In order to find the actual position of the particle at a given time, one has to integrate equation for $\dot{\varphi}$ above, of course.

The three first-order differential equations (I.46) together with the constant angular momentum $\vec{L}$ thus provide the complete solution for a particle's motion in an arbitrary central force field. The required input involves directly the force through the scalar force function $f(r)$. The corresponding orbit equation is explicitly expressed in terns of *LRL* vector $\vec{A}$ as a local conic section relative to $\vec{A}$, with the local eccentricity given by $\varepsilon(\varphi) = \left(A_1^2 + A_2^2\right)^{1/2}$. To solve these differential equations numerically and then determine the orbit is a rather trivial task with modern computer facilities.

Equations of Motion for Arbitrary Forces

Now, using the principles of above considered method, we derive the equation of motion for a particle moving in an arbitrary force field. The strategy will be such that of the required six first-order equations, three will be given through the Cartesian components of the LRL vector, two through the angles describing the orientation of the position $\vec{r}$, and one through a suitable chosen component of the angular momentum.

In spherical polar coordinates, where r is the radial distance, and θ and φ are the polar and asimuthal angles, respectively, of the position vector $\vec{r} = (r, \vartheta, \varphi)$ and where $\vec{e}_{r,} \vec{e}_\vartheta$ and $\vec{e}_\varphi$ are the corresponding unit vectors, an arbitrary force $\vec{F}(\vec{r})$ may be decomposed as

$$\vec{F}(\vec{r}) - F_r\vec{e}_r + F_\vartheta\vec{e}_\vartheta + F_\varphi\vec{e}_\varphi \qquad \text{(I.48)}$$

The analogous decomposition for the angular momentum yields

$$\vec{L} = l_\varphi\vec{e}_\varphi - l\,\vec{e}_\vartheta \qquad \text{(I.49)}$$

with the components

$$l_\varphi = mr^2\dot{\theta}$$
$$l = mr^2\dot{\varphi}\sin\theta$$

Note that the sign of the θ component is chosen such that for those problems in which motion is confined to a plane (taken as the (xy)-plane), it coincides with the angular momentum defined in the previous example.

Leaving the definition of the LRL vector unchanged, its decomposition is then found to be

$$\vec{A} = \left(1 - \frac{l^2}{mar}\right)\vec{e}_r + \left(\frac{l_\varphi}{a}\right)\vec{e}_\vartheta + \left(\frac{l\dot{r}}{a}\right)\vec{e}_\varphi \qquad \text{(I.50)}$$

with $\quad L^2 = \vec{L}\cdot\vec{L} = l_\varphi{}^2 + l^2 \qquad$ (I.51)

Taking the relevant scalar products, we find

$$\dot{r} = \frac{a}{l}\vec{A}\cdot\vec{e}_\varphi = \frac{a}{l_\varphi}\vec{A}\cdot\vec{e}_\vartheta \qquad \text{(I.52)}$$

The last equality, which is also a consequence of $\vec{A}\cdot\vec{L}=0$, allows us to eliminate one component of the angular momentum in favor of the other; we choose to eliminate l_φ .Hence, introducing

$$\beta=\frac{\vec{A}\cdot\vec{e}_\vartheta}{\vec{A}\cdot\vec{e}_\varphi} \tag{I.53}$$

We have

$$l_\varphi=\beta\cdot l$$

and consequently

$$L^2=l^2\left(1+\beta^2\right) \tag{I.54}$$

As earlier, the scalar product of $\vec{A}$ and $\vec{e}_r$ yields the orbit equation, we again find the familiar form

$$r=\frac{L^2}{ma}\frac{1}{1-\vec{A}\cdot\vec{e}_r} \tag{I.55}$$

which shows once more that locally all orbits can be interpreted as conic sections relative to the changing LRL vector.

What remains to be done now is to actually write down the six first-order equations of motion for the determination of the angles φ and ϑ, the angular momentum component l, and the three Cartesian components of $\vec{A}=\left(A_1,A_2,A_3\right)$. The first three follow from Newton's equation of motion in spherical polar coordinates and read

$$\dot{\varphi}=\frac{l}{mr^2\sin\vartheta}$$

$$\dot{\vartheta}=\beta\frac{l}{mr^2}$$

$$\dot{l}=rF_\varphi-\beta\frac{l^2}{mr^2}\cos\vartheta$$

From the decomposition of $\vec{A}$ it follows:

$$A_1=A\sin\vartheta\cos\varphi+B\cos\vartheta\cos\varphi-C\sin\varphi$$

$$A_2 = A\sin\vartheta\sin\varphi + B\cos\vartheta\sin\varphi + C\cos\varphi$$
$$A_3 = A\cos\vartheta - B\sin\vartheta$$

where the coefficients are

$$A = -\frac{2l}{ma}\left(F_\varphi + \beta F_\vartheta\right)$$
$$B = \frac{L}{ma}\left[\beta\left(\frac{a}{r^2} + F_r\right) + \left(1+\beta^2\right)\frac{\vec{A}\cdot\vec{e}_\varphi}{1-\vec{A}\cdot\vec{e}_\varphi}F_\vartheta\right]$$
$$C = \frac{L}{ma}\left[\frac{a}{r^2} + F_r + \left(1+\beta^2\right)\frac{\vec{A}\cdot\vec{e}_\varphi}{1-\vec{A}\cdot\vec{e}_\varphi}F_\varphi\right]$$

With the help the orbit equation (I.55), the solutions of above the six first-order differential equations thus provide the desired result: a complete description of the trajectory $\vec{r}(t) = \left(x(t), y(t), z(t)\right)$ of a particle under the influence of an arbitrary force.

Summary Comments on Dynamical Symmetries in Classical (Non-Relativistic) Mechanics

In this Chapter main attention was drawn to the relation between symmetry and a degeneracy of motion in non-relativistic classical mechanics.

We saw that the additional symmetries are mainly ascribed to Bertrand's theorem stating that the Kepler problem and the isotropic harmonic oscillator are only central potentials having closed non precessing orbits. The restrictions of this theorem emphasize the special role of the LRL vector. It is based on a global symmetry generated by vectorial integrals of motion which are continuous single-valued vectors in space.

When switching to a local viewpoint considering only local symmetries of a problem one can allow multi-valued constants of motion, as usual. But $O(4)$ symmetry eliminates the special role of the abovementioned central force problems.

As examples for studies such diverse problems of weakly broken symmetry will be considered in the following Sections. It'll be shown that for every bounded trajectory of a particle in an arbitrary central field a uniformly rotating reference frame can be chosen in which this trajectory becomes closed.

Algebraic approach to non-relativistic classical motion in central potentials based on $O(4)$ as local symmetry group established generalized LRL vector in the form

$$\vec{A} = f_1(r)\hat{\vec{r}} + f_2(r)\vec{L}\times\vec{p}$$

as suitable instruments to describe the motion. The unknown functions $f_i(r)$ have to be adapted to the problem at hand by requiring for this generalized vector analogous properties as exhibited for original LRL vector. These functions depend in general on the energy, angular momentum and the mass as parameters.

The symmetry is called *local* since the orbit can be considered locally as a conic section whose eccentricity is determined by the magnitude of the time- depedent LRL vector. The symmetry is weakly (minimally) broken in the sense that coefficient functions $f_i(r)$ can become singular at the turning points and the vector $\vec{A}$ keeps it length constant while rotating with constant angular speed.

In the following Chapter we'll see first of all that this concept of minimally broken symmetry remains valid in case of post-Newtonian approximation which suggests that some traces of hidden symmetry should be manifest themselves in the relativistic case.

REFERENCES

[1] Goldstein H., Poole C., Saiko J., *Classical Mechanics,* 3rd Ed. Pearson Educ. Inc, 2004.

[2] McCauley J.L., *Classical Mechanics,* Cambridge, 1998.

[3] Tikochinsky Y. *A Simplified Proof of Bertrand's Theorem, Am.J.Phys.* 56, 1073 (1988).

[4] Zarmi Y. *The Bertrand Theorem Revisited, Am.J.Phys.* 70, 446 (2002).

[5] Goldstein H.,*Prehistory of the Runge-Lenz Vector,* Am.J.Phys. 43,1737 (1975)

[6] Goldstein H., *More on Prehistory of the Laplace-Runge-Lenz Vector", Am.J.Phys.* 44,1123 (1976).

[7] Popov V.S., *On Hidden Siymmetry of the Hydrogen Atom",* In Proceedings of Int. School "Physics of High Energy and Elementary Particles", *Naukova Dumka,* Kiev, 1967, p.702.

[8] Dulock V.A. McIntosh H.V., *On the Degeneracy of the two-Dimensional Harmonic Oscillator, Am.J.Phys.* 33,109 (1965)

[9] Fradkin D.M., *Three Dimensional Isotropic Harmonic Oscillator and SU(3), Am.J.Phys.* 39, 207 (1965).

[10] Fradkin D.M., *Existence of the Dynamical Symmetries O(4) and SU(3) for all Classical Central Potential Problems, Progr. Theor. Phys.,* 37, 798 (1967)

[11] Serebrennikov V.B. Shabad A.E., *Method of Calculation of the Radial-Symmetric Hamiltonian Spectrum on the Basis of Approximated O(4) and SU(3)-Symmetry, Theor. Math.Phys.,*8,644 (1971).

[12] Halas A., March N.H., *A Generalisation of the Runge-Lenz Constant of Classical Motion in a Central Potential, J.Phys.A: Math. Gen.* 23, 735 (1990).

[13] Bartnik et al., *Equation of Motion Using the Dynamical Evolution of the Runge-Lenz Vector. Astrophys. Journ.* 334, 517 (1988).

[14] Truax D.R., *Dynamical Symmetries of Rotationally Invariant, Three Dimensional, Schroedinger Equation. J.Math.Phys.* 21,807 (1980).

Chapter II

HIDDEN SYMMETRY IN CLASSICAL RELATIVISTIC MECHANICS

"He (Galileo) held that Science and Religion were not enemies, but rather allies –two different languages telling the same story, a story of symmetry and balance."

Dan Brown

II.1. AUXILIARY PROBLEM: LRL VECTOR FOR A MODIFIED KEPLER PROBLEM

In studies of planetary motion small additional inverse cube force is also often considered, or potential of kind

$$V(r) = -\frac{a}{r} + \frac{\beta}{r^2}, \qquad (\alpha > 0, \quad \beta > 0) \text{ (II.1)}$$

Small additional terms in gravitational potential can be caused by small perturbations from other planets, or by deviation of sun's form from exact sphericity. These kinds of perturbations provide a slow precession of orbit with respect to center of mass of sun's system.

Remark also that an orbit precession may be caused by relativistic effects. For example, relativistic electron in hydrogen atom moves on rosette trajectories. Therefore, the consideration of this additional term in Kepler problem has a serious physical ground.

The effective potential in this problem looks like:

$$V_{eff}(r) = \frac{\widetilde{L}^2}{2mr^2} - \frac{a}{r} \tag{II.2}$$

where

$$\widetilde{L} = \gamma L, \quad \gamma = \sqrt{1 + \frac{2m\beta}{L^2}} \tag{II.3}$$

The solution for the trajectory is written now as

$$\gamma\, \theta = \int_{r_{\min}}^{r_{\max}} \frac{\widetilde{L} dr}{r^2 \sqrt{2mE + \frac{2ma}{r} - \frac{\widetilde{L}^2}{r^2}}}$$

We see from here that the orbit equation is

$$\frac{\widetilde{p}}{r} = 1 + \widetilde{\varepsilon} \cos \gamma\theta \tag{II.4}$$

where

$$\widetilde{p} = \frac{\gamma^2 L^2}{ma}, \qquad \widetilde{\varepsilon} = \sqrt{1 + \frac{2E\gamma^2 L^2}{ma^2}} \tag{II.5}$$

In spite of outward likeness to the solution for pure Kepler problem, trajectories diverse significantly and as more as γ is different from 1. This difference is particularly eminent in case of finite $(E < 0)$ motion. To see this let us calculate the angular displacement in the course of full period

$$\Delta\theta = 2 \int_{r_{\min}}^{r_{\max}} \frac{L dr}{r^2 \sqrt{2m\left(E - V_{eff}(r)\right)}} = \frac{2\pi}{\gamma} \tag{II.6}$$

The magnitude of γ is dependent on the parameters of the problem. Most often γ is irrational number and not so much different from 1 (when $2m\beta << L^2$). In this case the motion takes place on rosette trajectories – the body moves on elliptic orbit, which by itself is precessing slowly relative to center of field. The velocity of this precession is characterizing by quantity

$$\delta\theta = \Delta\theta - 2\pi = 2\pi\left(\frac{1}{\gamma} - 1\right)$$

When $2m\beta << L^2$, it follows approximately

$$\delta\theta \approx -\frac{2mn\beta}{L^2}$$

Let us remark that the similar precessional motion of elliptic orbits is observed also in the potential

$$V(r) = -\frac{a}{r} - \frac{\beta}{r^2}, \qquad (\alpha, \beta > 0) \tag{II.7}$$

Let consider this problem by Fradkin's method [1]. In this case

$$H = \frac{p^2}{2m} - \frac{a}{r} - \frac{\beta}{r^2} \tag{II.8}$$

and the orbit equation can be derived by integration:

$$\frac{\lambda}{r} = 1 + \varepsilon' \cos\gamma'\theta \tag{II.9}$$

where

$$\lambda = \frac{\gamma'^2 L^2}{ma}, \qquad \gamma' = \sqrt{1 - \frac{2m\beta}{L^2}} \tag{II.10}$$

and eccentricity is

$$\varepsilon' = \sqrt{1 + \frac{2\gamma'^2 L^2 E^2}{ma}} \tag{II.11}$$

The orbit is restricted to the noncircular bound motion, so that $0 < \varepsilon' < 1$.

Accordingly $0 > E > -\frac{ma}{2\gamma'^2 L^2}$.

Fradkin's constant vector for the Kepler – problem can in fact be constructed according to equations considered earlier. Namely

$$\hat{\vec{k}} = \left(r^2 \frac{\partial f}{\partial r} \right) \frac{\vec{p} \times \vec{L}}{L^2} + \left(\frac{f}{r} + \frac{\partial f}{\partial r} \right) \vec{r} \tag{II.12}$$

where a function f is defined as

$$f(r,L,E) \equiv \cos\theta = \cos\left(\int r^{-2} \left[\frac{2m(E-V)}{L^2} - r^{-2} \right]^{-\frac{1}{2}} dr \right) \quad \text{(II.13)}$$

However, on account of the character of orbit $\left(\frac{\lambda}{r}\right)$ the vector cannot be expressed by simple functions of the orbit.

Now let us introduce a rotating coordinate system, denoted by a prime, defined by rotation mutually orthogonal unit vectors [2],

$$\begin{aligned} \vec{i}\,' &= \vec{e}_r \cos\gamma'\theta - \vec{e}_\theta \sin\gamma'\theta = \\ &= +\vec{i}\cos(1-\gamma')\theta + \vec{j}\sin(1-\gamma')\theta \end{aligned}$$

$$\begin{aligned} \vec{j}\,' &= \vec{e}_r \sin\gamma'\theta + \vec{e}_\theta \cos\gamma'\theta = \\ &= -\vec{i}\sin(1-\gamma')\theta + \vec{j}\cos(1-\gamma')\theta \\ \vec{k}\,' &= \vec{k} \end{aligned}$$

The unit rotating vector $\vec{i}\,'$ is rewritten in terms of canonical variables $\vec{r}$ and $\vec{p}$ as

$$\vec{i}\,' = \left(r^2 \frac{\partial g}{\partial r} \right) \frac{\vec{p} \times \vec{L}}{\gamma' L^2} + \left(\frac{g}{r} + \frac{1}{\gamma'} \frac{\partial g}{\partial r} \right) \vec{r} \quad \text{(II.14)}$$

where g is

$$g(r,L,E) \equiv \cos\gamma'\theta$$

Hence a desired LRL vector is constructed as

$$\vec{A}_R = \gamma'\,\vec{p} \times \vec{L} - \left[(1-\gamma')\gamma' L^2 + mar \right] \frac{\vec{r}}{r^2} = ma\varepsilon'\,\vec{i}\,' \quad \text{(II.15)}$$

If $\gamma' = 1$, the above equation is equal to the LRL vector.

The orbit equation can be found as usual

$$\vec{A}_R \cdot \vec{r} = ma\varepsilon'\,\vec{i}\,' \cdot \vec{r} = ma\varepsilon' r \cos\gamma'\theta$$

On the other hand, with the aid of $\vec{r} \cdot (\vec{p} \times \vec{L}) = L^2$, one obtains

$$\vec{A}_R \cdot \vec{r} = \gamma'^2 L^2 - mar$$

Therefore the orbit is given by

$$\frac{\gamma'^2 L^2}{mar} = 1 + \varepsilon' \cos\gamma'\theta \tag{II.16}$$

which is nothing but the original equation.

Here we shall sketch the relation of dynamical variables in coordinates K and K', while K is the fixed system defined by $(\vec{i}, \vec{j}, \vec{k})$ and K' is the rotating system defined by $(\vec{i}', \vec{j}', \vec{k}')$.

The position vectors are the same in both systems

$$\vec{r} = x\vec{i} + y\vec{j} = x'\vec{i}' + y'\vec{j}' = \vec{r}'$$

The components of $\vec{r}'$ are

$$x' = x\cos(1-\gamma')\theta + y\sin(1-\gamma')\theta$$
$$y' = -x\sin(1-\gamma')\theta + y\cos(1-\gamma')\theta$$

The linear momentum $\vec{p}'$ in the system K' is defined as

$$\vec{p}' = p'_x\vec{i}' + p'_y\vec{j}'$$

so that one gets

$$\begin{aligned}\frac{\vec{p}'}{m} &= \frac{dx'}{dt}\vec{i}' + \frac{dy'}{dt}\vec{j}' = \\ &= \left[\frac{dx}{dt} + (1-\gamma')\frac{d\theta}{dt}y\right]\vec{i} + \left[\frac{dy}{dt} - (1-\gamma')\frac{d\theta}{dt}x\right]\vec{j} = \\ &= \frac{\vec{p}}{m} - (1-\gamma')\frac{d\theta}{dt}\vec{k}\times\vec{r}\end{aligned}$$

Thus the system K' is rotating for the fixed system K with angular velocity

$$\vec{\Omega} = (1-\gamma')\frac{d\theta}{dt}\vec{k} \tag{II.17}$$

Using $\vec{L} = mr^2\frac{d\theta}{dt}\vec{k}$, one obtains

$$\vec{\Omega} = (1-\gamma')\frac{\vec{L}}{mr^2} \tag{II.18}$$

Moreover it is shown that the vector $\vec{A}_R$ also rotates with the same angular velocity as follows.

$$\vec{A}_R \times \{A_R, H\} / A_R^2 = \vec{\Omega}$$

where braces express the Poisson bracket by canonical variables.

Now let us show that using vector $\vec{\Omega}$ the secular precession rate of this problem is calculated. Integrating this vector $\vec{\Omega}$ over a time interval τ, we have

$$\langle\vec{\Omega}\rangle \equiv \frac{1}{\tau}\int_0^\tau \vec{\Omega}dt = \frac{1}{\tau}(1-\gamma')\vec{k}\int_0^\tau \frac{d\theta}{dt}dt$$

where τ is a single orbital period of the motion. Then we can interchange the range of integration from the time $0 \le t \le \tau$ to the angle $0 \le \theta \le \dfrac{2\pi}{\gamma'}$, so that one gets

$$\langle\vec{\Omega}\rangle = 2\pi\left(\frac{1}{r'} - 1\right)\frac{\vec{k}}{r} \tag{II.19}$$

This is exactly the same as secular precession of the orbit given by ordinary method above.

From the relation of momentum of two coordinate systems one easily finds

$$p'^2 = p^2 = (1-\gamma'^2)\frac{L^2}{r^2} = p^2 - \frac{2m\beta}{r^2}$$

and angular momentum

$$\vec{L}' = \vec{r}' \times \vec{p}' = \gamma'\vec{L}$$

So that

$$L'^2 = L^2 - 2m\beta$$

Therefore the Hamiltonian H and rotating LRL vector $\vec{A}_R$ are rewritten in terms of primed quantities, namely, we have

$$H = \frac{p'^2}{2m} - \frac{a}{r'}$$

which is the same form as the Kepler-Coulomb problem, and rotating LRL vector in the system K' is

$$A'_R = \gamma'\left[\vec{p}' + (1-\gamma')\frac{\vec{L}\times\vec{r}}{r^2}\right]\frac{\vec{L}'}{\gamma'} - \left[(1-\gamma')\gamma' L^2 + \frac{ma}{r'}\right]\frac{\vec{r}'}{r'^2} = \vec{p}'\times\vec{L}' - ma\frac{\vec{r}'}{r} \quad \text{(II.20)}$$

which is nothing but the LRL vector in the fixed system K.

We can conclude that in the rotating system K' which rotates with $\vec{\Omega}$, the modified problem is reduced to the pure Kepler-Coulomb problem, that is, the inverse cube force vanishes and the turning point of precessing orbit is fixed.

The orbit is closed and single-valued in the rotating system but open and infinitely many valued in the inertial system.

In future we will see that the relativistic Kepler problem has this peculiarity.

II.2. The Laplace-Runge-Lenz Vector and the Lorentz Boost

As we saw in the previous Chapter the dynamical symmetry of the non-relativistic Kepler problem manifests itself through the LRL vector, which is a constant of motion. It is more relevant and important to ask a question: Does the LRL vector have a deeper physical origin?

Now we shall show that it is possible to tie the dynamical symmetry of the non-relativistic Kepler problem to the classical invariance principles of special relativity [3]. For that aim it is necessary to investigate the relativistic *two-body problem* in order to make clear the connection between the dynamical symmetry of the non-relativistic Kepler problem and special relativity.

The invariance group of electromagnetic (gravitational) many-body problem is the inhomogeneous Lorentz group. It is well known that this group forces ten constants of the motion upon a physical system – the generators of the infinitesimal transformations of the group. The total momentum $\vec{P}$ and the total energy $H = E$ generate translations in space and time, and they form together a four-vector $\left(\frac{E}{c}, \vec{P}\right)$. The generators for the homogeneous Lorentz group are the total angular momentum $\vec{L}$, generating rotations in three-space and a polar vector $\vec{K}$ which generates homogeneous Lorentz thansformations without ordinary rotations, called Lorentz boosts.

Our aim will be to show that the existence of conserved LRL vector is a simple consequence of the fact that $\vec{K}$ is a constant of the motion.

Post-Newtonian approximation means that one has to consider a Lagrangian for a system of N-partcle in $1/c^2$ order. This was done by Darwin [4] and is known as the Darwin Lagrangian, which for particles with rest masses $m_1, m_2, ..., m_N$ and charges $q_1, q_2, ..., q_N$ has the following form [4, 5] :

$$L = -\sum_i m_i c^2 + \frac{1}{2}\sum_i m_i \upsilon_i^2 + \frac{1}{8c^2}\sum_i m_i \upsilon_i^4 - \frac{1}{2}\sum_{i\neq j}\frac{q_i q_j}{r_{ij}} + \frac{1}{4c^2}\sum_{i\neq j} q_i q_j \left[\frac{\vec{\upsilon}_i \vec{\upsilon}_j}{r_{ij}} + \frac{(\vec{\upsilon}_i \vec{r}_{ij})(\vec{\upsilon}_j \vec{r}_{ij})}{r_{ij}^3}\right]$$

where $\vec{\upsilon}_i$ is the velocity of the i-th particle, with position vector $\vec{r}_{ij}$, and

$$\vec{r}_{ij} = \vec{r}_i - \vec{r}_j, \quad r_{ij} = \left|\vec{r}_{ij}\right|.$$

We get from this Lagrangian

$$\vec{p}_i = \frac{\partial L}{\partial \vec{\upsilon}_i} = m_i \vec{\upsilon}_i + \frac{m_i \upsilon_i^2}{2c^2}\vec{\upsilon}_i + \frac{1}{2c^2}\sum_{j\neq i} q_i q_j \left(\frac{\vec{\upsilon}_j}{r_{ij}} + \frac{\vec{\upsilon}_j \vec{r}_{ij}}{r_{ij}^3}\vec{r}_{ij}\right) \qquad \text{(II.21)}$$

The Hamiltonian or energy is constructed as usual

$$H = \sum_i \vec{p}_i \vec{\upsilon}_i - L$$

and, to order $1/c^2$, becomes

$$H = \sum_i m_i c^2 + \sum_i \frac{p_i^2}{2m_i} - \frac{1}{c^2}\sum_i \frac{p_i^4}{8m_i^3} + \frac{1}{2}\sum_{j\neq i}\frac{q_i q_j}{r_{ij}} - \frac{1}{4c^2}\sum_{j\neq i}\frac{q_i q_j}{m_i m_j}\left[\frac{\vec{p}_i \vec{p}_j}{r_{ij}} + \frac{(\vec{p}_i \vec{r}_{ij})(\vec{p}_i \vec{r}_{ij})}{r_{ij}^3}\right]$$

For the total momentum and total angular momentum we have

$$\vec{P} = \sum_i \vec{p}_i \quad \text{and} \quad \vec{L} = \sum_i \vec{r}_i \times \vec{p}_i$$

The corresponding expression for the boost vector $\vec{K}$ we extract from the Fock's book [5]

$$\vec{K} = -t\vec{P} + \sum_i \left[m_i \vec{r}_i + \frac{1}{2c^2}\left(\frac{p_i^2}{m_i} + \sum_{j\neq i}\frac{q_i q_j}{r_{ij}}\right)\vec{r}_i\right] \qquad \text{(II.22)}$$

Now one can to introduce the vector of the centre of inercia $\vec{R}_C$, by the following relation

$$E\vec{R}_C = \sum_i \mathrm{E}_i \vec{r}_i$$

where E_i is the energy associated with the ith particle

$$\mathrm{E}_i = m_i c^2 + \frac{p_i^2}{m_i} + \frac{1}{2}\sum_{j\neq i}\frac{q_i q_j}{r_{ij}} \tag{II.23}$$

Then the following relation follows

$$\vec{K} = \frac{\mathrm{E}}{c^2}\vec{R}_C - t\vec{P} \tag{II.24}$$

which shows that $\vec{K}$ is a constant vector of the centre of inercia. This vector depends explicitly on time, except in the center of momentum system which is the Lorentz frame for which $\vec{P}=0$. In this particular reference system $\vec{K}$ becomes a time independent constant of the motion.

In particular case of two-body system, we have

$$\vec{K} = (m_1 + m_2)\vec{R} + \frac{1}{c^2}\left(\frac{p^2}{2\mu} + \frac{q_1 q_2}{r}\right)\vec{R} - \frac{1}{2c^2}\frac{m_1 - m_2}{m_1 + m_2}\left(\frac{p^2}{\mu} + \frac{q_1 q_2}{r}\right)\vec{r} \tag{II.25}$$

Here μ is the reduced mass $\mu = \dfrac{m_1 m_2}{m_1 + m_2}$, while $\vec{R} = \dfrac{m_1\vec{r}_1 + m_2\vec{r}_2}{m_1 + m_2}$ is the position vector of the centre of mass and $\vec{r}$ is a relative momentum vector.

For a two-body problem the Hamiltonian becomes

$$\begin{aligned} H = (m_1 + m_2)c^2 + \frac{p^2}{2\mu} + \frac{q_1 q_2}{r} - \frac{1}{8c^2}\left(\frac{1}{m_1^3} + \frac{1}{m_2^3}\right)p^4 + \\ + \frac{q_1 q_2}{2c^2}\frac{1}{m_1 m_2}\left(\frac{p^2}{r} + \frac{(\vec{p}\vec{r})^2}{r^3}\right) \end{aligned} \tag{II.26}$$

In order $1/c^2$ we have

$$\frac{H}{c^2} = m_1 + m_2 + \frac{1}{c^2} H_{NR} \tag{II.27}$$

where

$$H_{NR} = \frac{p^2}{2\mu} + \frac{q_1 q_2}{r} \tag{II.28}$$

is the non-relativistic Hamiltonian. Hence

$$\vec{K} = \frac{1}{c^2} H\vec{R} - \frac{1}{2c^2} \frac{m_1 - m_2}{m_1 + m_2} H_{NR}\vec{r} \tag{II.29}$$

In the centre of momentum system $\vec{P}$ is no longer a dynamical variable. The same holds, therefore, for the corresponding position vector $\vec{R}$.

The expression for $\vec{K}$ contains dependence on $\vec{R}$. It must be eliminated from this expression in order for $\vec{K}$ to be a proper dynamical function.

With H being the Hamiltonian, Hamilton's equation for $\vec{R}$ gives

$$\frac{d\vec{R}}{dt} = \left(\frac{\partial H}{\partial \vec{P}}\right)_{\vec{P}=0}$$

or according to definitions

$$\frac{d\vec{R}}{dt} = \left(\frac{m_1}{m_1 + m_2} \frac{\partial H}{\partial \vec{p}_1} + \frac{m_2}{m_1 + m_2} \frac{\partial H}{\partial \vec{p}_2}\right)_{\substack{\vec{p}_1 = \vec{p} \\ \vec{p}_2 = -\vec{p}}}$$

which becomes

$$\frac{d\vec{R}}{dt} = \frac{1}{2c^2} \frac{m_1 - m_2}{(m_1 + m_2)^2} \frac{1}{\mu} \left[H_{NR}\vec{p} + \frac{q_1 q_2}{r^3} \vec{r}(\vec{p}\vec{r}) \right]$$

In order $1/c^2$ we have

$$\frac{d\vec{R}}{dt} = \frac{1}{2c^2} \frac{m_1 - m_2}{(m_1 + m_2)^2} \frac{1}{\mu} \frac{d}{dr} [(\vec{p}\vec{r})\vec{p}]$$

From which we get by integration

$$\vec{R} = \vec{R}_0 + \frac{1}{2c^2}\frac{m_1 - m_2}{(m_1 + m_2)^2}\frac{1}{\mu}(\vec{p}\vec{r})\vec{p} \tag{II.30}$$

where $\vec{R}_0$ is an arbitrary vector. It defines the position of center of mass at a time $(\vec{p}\vec{r})/c^2 = 0$. Each conic section possesses at least one such point and the last expression offers, therefore, a well-defined correlation between the orbital motion and the motion of the center of mass for any electromagnetic two-body system. We need such a correlation in order to set up a well-defined expression for the constant of motion defined by $\vec{K}$.

Inserting $\vec{R}$ into $\vec{K}$, we obtain

$$\vec{K} = \frac{1}{c^2}H\vec{R}_0 - \frac{1}{2c^2}\frac{m_1 - m_2}{m_1 + m_2}\vec{A}$$

where

$$\vec{A} = \left(\frac{p^2}{\mu} + \frac{q_1 q_2}{r}\right)\vec{r} - \frac{1}{\mu}(\vec{p}\vec{r})\vec{p} = \frac{\vec{p}\times\vec{L}}{\mu} + \frac{q_1 q_2}{r}\vec{r} \tag{II.31}$$

It is a LRL vector! We have thus reached our goal of demonstrating that the form of $\vec{A}$ is determined by the boost vector $\vec{K}$ in the center-of-momentum (CM) system.

The fact that $\vec{K}$ is a constant of motion implies that the Poisson bracket between H and $\vec{K}$ must vanish in the CM system:

$$\{H, \vec{K}\} = 0$$

With $\vec{K}$ given by above expression and H/c^2, we find that

$$\{H, \vec{K}\} = -\frac{1}{2c^2}\frac{m_1 - m_2}{m_1 + m_2}\{H, \vec{A}\} \tag{II.32}$$

Relation $\{H, \vec{K}\} = 0$ obviously implies that $\{H, \vec{A}\} = 0$, i.e. the LRL vector must commute with the non-relativistic Hamiltonian. So, we have shown that the vanishing of Poisson bracket of $\vec{A}$ with Hamiltonian is a consequence of the vanishing of that of $\vec{K}$ with Hamiltonian.

The same conclusion remains valid, naturally, in the problem of the gravitating bodies.

Therefore, we have tied the LRL vector to the generator $\vec{K}$ of the infinitesimal Lorentz transformations. This connection reflects, in turn, the requirement that the dynamics must be in accordance with the structure of the inhomogeneous Lorentz group. It is a strong requirement on the form of dynamics, as thoroughly discussed by Dirac [6]. Dirac refers, in fact, jointly to H and $\vec{K}$ as the Hamiltonians, thus stressing that the forms of H and $\vec{K}$ are strongly coupled by the group requirements. So strong is this coupling that its implications are also felt in the non-relativistic limit, namely through the requirement that $\vec{A}$ must exist as a constant of motion in that limit.

II.3. Post-Newtonian Extensions of the LRL Vector

The basis question arised from the previous result is the following: Given post-Newtonian equations of motion that correspond to a two-body system, does there exists any post-Newtonian extension of the LRL vector [7]?

Now, we prove that such a post-Newtonian extension is possible for a very large class of systems which include the relevant physical interactions (electromagnetism and gravitation).

Let us consider a relativistic two-body system, constituted by two pointlike particles, whose equations of motion at the post-Newtonian level are obtained from the Hamiltonian in the CM system (see previus Section):

$$H(r,(\vec{r}\vec{p}),\vec{p})=\frac{p^2}{2\mu}-\frac{\kappa}{r}+\frac{1}{c^2}\left\{a\frac{p^4}{\mu^3}-\frac{\kappa}{r}\left[b\frac{p^2}{\mu^2}-d\frac{\kappa}{\mu r}+e\frac{(\vec{r}\vec{p})^2}{r^2\mu^2}\right]\right\} \qquad \text{(II.33)}$$

It is rewritten in this form in order to unify two cases under consideration.

Here κ is a constant, $\vec{p}$ is the relative momentum $\left[i.e.\ \vec{p}=\mu\vec{\upsilon}+O(c^{-2})\right]$ and a,b,c,d,e are arbitrary functions of dimensionless variable $\mu(m_1+m_2)^{-1}$. This structure corresponds to a nonsingular (except at $r=0$) relativistic extension, invariant under time translations and spatial rotations, of a classical Hamiltonian with a Coulomb potential.

This expression coincides in the CM frame with Darwin's Hamiltonian (DH) [6] and Einstein-Infeld-Hoffmann's Hamiltonian (EIHH) [8] for the particular values:

$$\text{DH:} \quad a=\frac{1}{8}\left(-1+\frac{3\mu}{m}\right),\quad b=\frac{\mu}{2m},\quad d=0,\quad e=\frac{\mu}{2m},\quad \kappa=-e_1e_2,\quad m=m_1+m_2$$

$$\text{EIHH:}\quad a=\frac{1}{8}\left(-1+\frac{3\mu}{m}\right),\quad b=\frac{1}{2}\left(3+\frac{\mu}{m}\right),\quad d=\frac{1}{2},\quad e=\frac{\mu}{2m},\quad \kappa=Dm_1m_2$$

This Hamiltonian describes the electromagnetic (gravitational) interaction of two pointlike particles up to the post-Newtonian level in the framework of classical electrodynamics (general relativity).

We shell look for a post-Newtonian extension of LRL vector $\vec{A}_N \equiv \vec{\upsilon} \times \vec{L}_N - \kappa\vec{r}/r$, orthogonal to the relative post-Newtonian angular momentum $\vec{L}_N = \vec{r} \times \vec{p}$, being a vector first integral of the equation of motion.

We obtain easily

$$\vec{p} = \mu\vec{\upsilon} + \frac{1}{c^2}\left\{-4a\mu\upsilon^2\vec{\upsilon} + \frac{2\kappa}{r}\left[b\vec{r} + e\frac{\vec{r}\vec{\upsilon}}{r^2}\vec{r}\right]\right\} \qquad \text{(II.34)}$$

$$\vec{L} = \left\{1 + \frac{1}{c^2}\left[-4a\upsilon^2 + 2b\frac{\kappa}{\mu r}\right]\right\}\vec{L}_N \qquad \text{(II.35)}$$

Then, we shell consider the decomposition

$$\vec{A} = \vec{A}_N + \frac{1}{c^2}\left[f\vec{A}_N + g\vec{L} \times \vec{A}_N\right] \qquad \text{(II.36)}$$

Our objective is to calculate the functions f and g, which are assumed invariant under spatial rotations. They must be arbitrary functions of three independent scalars, for instance $\{r, E_N, L_N\}$, where $L_N = \left|\vec{L}_N\right|$.

Here $\vec{A}$ is a vector first integral if its Poisson bracket with Hamiltonian vanishes.

First of all let us rewrite $\vec{A}$ in the form

$$\vec{A} = \vec{A}_0 + \frac{1}{c^3}\left[\alpha\vec{A}_0 + \beta\vec{L} \times A_0\right],\ \alpha, \beta = F(r, L, E) \qquad \text{(II.37)}$$

with

$$\vec{A}_0 = \left(\frac{p^2}{\mu} - \frac{\kappa}{r}\right)\vec{r} - \frac{(\vec{r}\vec{p})}{\mu}\vec{p} \qquad \text{(II.38)}$$

After calculating the Poisson bracket of this $\vec{A}$ with H derives a couple of partial differential equations with respect to r for α and β, whose integrations give [7]

$$\alpha = \frac{L^2}{A_0^2}\frac{\kappa}{\mu r}\left[2(4a - b - e)\frac{E}{\mu} + (4a - 2b + d - 2e)\frac{\kappa}{\mu r} + e\left(\frac{L}{\mu r}\right)^2\right] + \tilde{\alpha}(L, E) \qquad \text{(II.39)}$$

$$\beta = \frac{1}{A_0^2}\left\{(4a-2b+d-e)\frac{\kappa}{L}\left(2\frac{E}{\mu}+\frac{\kappa^2}{L^2}\right)\tan^{-1}\left(\frac{\mu\kappa r - L^2}{L(\vec{r}\vec{p})}\right)+\right\}$$

$$+\frac{1}{A_0^2}\frac{\kappa(\vec{r}\vec{p})}{\mu r}\left[2(4a-b-e)\frac{E}{\mu}+(4a-2b+d-e)\left(\frac{\kappa^2}{L^2}+\frac{\kappa}{\mu r}\right)+e\left(\frac{L}{\mu r}\right)^2\right]+\widetilde{\beta}(L,E) \quad \text{(II.40)}$$

where $\widetilde{\alpha}, \widetilde{\beta}$ are arbitrary functions of their arguments, and $A_0 = \left|\vec{A}_0\right|$.

Reintroducing the value of $\vec{p}$ given above, we derive

$$f = \frac{L_N^2}{A_N^2}\frac{\kappa}{\mu r}\left[2(-20a+3b+e)\frac{E_N}{\mu}+(-12a+2b+2e+d)\frac{\kappa}{\mu r}-e\left(\frac{L_N}{\mu r}\right)^2\right]+\alpha^*(L_N, E_N)$$

$$g = \frac{1}{A_N^2}\left\{(4a-2b+d-e)\frac{\kappa^2}{L_N}\left(2\frac{E_N}{\mu}+\frac{\kappa^2}{L_N^2}\right)\tan^{-1}\left(\frac{\mu\kappa r - L_N^2}{\mu L_N(\vec{r}\vec{\upsilon})}\right)+\right. \quad \text{(II.41)}$$

$$\left.+\frac{\kappa}{r}(\vec{r}\vec{\upsilon})\left[(4a-2b+d-e)\frac{\kappa^2}{L_N^2}+(-12a+2b+d+e)\frac{\kappa}{\mu r}-e\left(\frac{L_N}{\mu r}\right)^2\right]\right\}+\beta^*(L_N, E_N)$$

where α^*, β^* are arbitrary functions of their arguments and $A_N = \left|\vec{A}_N\right|$..

In order to fix this arbitrariness, one must use the equation of orbit and fix the modulus of $\vec{A}$.

If one requires that the modulus of $\vec{A}$ depends on E and L in the same form as in Newtonian mechanics, i.e.

$$A^2 = \kappa^2 + \frac{2EL^2}{\mu}$$

we obtain the following expression

$$A^2 = \kappa^2 + \frac{2EL^2}{\mu} + \frac{2}{c^2}\left[A^2\alpha^* + 26a\left(\frac{EL}{\mu}\right)^2\right] \quad \text{(II.42)}$$

where from the expansion of these two expression, it is found that

$$\alpha^* = -26a\left(\frac{E}{\mu}\right)\left(2 + \frac{\mu\kappa^2}{EL^2}\right)^{-1}$$

and β^* is reduced to

$$\beta^* = \frac{\pi\kappa^2}{2L^3}$$

Taking into account all obtained relations, one can to construct a unique post -Newtonian extension of LRL vector which has a rather long expression, but it is possible to derive by it the usual Poisson algebra relations.

$$\begin{aligned} \{L_i, L_j\} &= \varepsilon_{ijk} L_k, \\ \{L_i, A_j\} &= \varepsilon_{ijk} A_k \\ \{A_i, A_j\} &= -\frac{2E}{\mu}\varepsilon_{ijk} L_k \end{aligned} \tag{II.43}$$

Therefore the Poisson algebra corresponding to the first integrals (L_i, A_i) is the same as in Newtonian mechanics, but now E appears instead of the Newtonian energy.

It remains as an unsolved problem the question of a possible extension at a higher order than the post-Newtonian level [7].

II.4. Relativistic Kepler Problem

Let us now consider the relativistic Kepler problem. We write the equation of motion as

$$m_0 \frac{d}{dt}\frac{\vec{\upsilon}}{\sqrt{1-\dfrac{\upsilon^2}{c^2}}} = -\frac{a}{r}\hat{\vec{r}} \tag{II.44}$$

It is suitable rewrite this equation in the frame defined by a transition from the time t to the proper time τ [9] accomplished by

$$d\tau = \sqrt{1-\upsilon^2/c^2}\,dt$$

Then this equation reads as

$$m_0 \frac{d^2\vec{r}}{d\tau^2} = -\frac{1}{m_0 c^2}\left\{E_{rel}\frac{a}{r^2} - \frac{a^2}{r^3}\right\}\hat{\vec{r}} \qquad \text{(II.45)}$$

where $E_{rel} = \sqrt{p^2c^2 + m_0^2c^4}$. Using τ formally as time one can interpret this equation of motion as a non-relativistic Kepler problem modified by an inverse cube force, i.e. as a perturbed Kepler problem with precessing orbits. As noted before this problem is most easily solved by a transition to a rotating coordinate system chosen such that the orbits appear closed. If we return to r^{-3} problem considered above the corresponding angular velocity of rotating system be

$$\vec{\Omega} = (1-\gamma')\frac{\vec{L}}{m_0 r^2} \qquad \text{(II.46)}$$

where

$$\gamma' = \sqrt{1 - \frac{2m_0\beta}{L^2}} \quad \text{and} \quad \beta = \frac{a^2}{2} > 0$$

Physically, the effect of this transition is to balance the inverse cube term by the so-called fictitious forces in the rotating system. This transition, as we already known, allows again a determination of the orbit via the scalar product of the LRL vector with $\vec{r}$. The straightforward transformation to the inertial system leads to the precessing orbit.

In a moving reference frame, such that the perihelion appears fixed, the orbit is non-relativistic in form. The corresponding transition is often called as Sommerfeld's transformation.

It is now easy to derive Sommerfeld's famous formula by applying the Bohr-Sommerfeld quantization rules to the phase integrals for p_r and p_φ,which are the usual radial momentum (p_r) and angular momentum (p_φ) in the decomposition

$$\vec{p}^2 = p_r^2 + \frac{1}{r^2}p_\varphi^2$$

The quantization rules look like

$$\int_{\varphi=0}^{\varphi=2\pi} p_\varphi d\varphi = n_\varphi \hbar$$

which yields immediately $p_\varphi = n_\varphi \hbar$, and

$$\int_{\varphi=0}^{\varphi=2\pi} p_r dr = n_r \hbar$$

It can be demonstrated that the sole effect of relativity in the radial quantization condition consists in replacing the orbital angular momentum $n_\varphi \hbar$ by the "irrational momentum" parameter $\gamma' n_\varphi \hbar$.

After some algebraic calculations the following result follows for the energy levels [9] :

$$E = m_0 c^2 \left[1 + \frac{a^2}{\left(n_r + \sqrt{n_\varphi^2 - a^2} \right)} \right]^{-1/2}$$

which is the well-known Sommerfeld's formula.

It would be very interesting to know how group-theoretical methods can be invented to explain the relativistic splitting of energetic levels in quantum mechanics.

References

[1] Fradkin D.M. Existence of the Dynamical Symmetries O(4) and SU(3) for All Classical Central Problems, *Progr. Theor. Phys.* 37, 98 (1967).

[2] Yoshida T. Considerations on the Precessing Orbit via a Rotating Laplace- Runge-Lenz Vector, *Am. J. Phys.* 55,1133 (1987).

[3] Dahl J.P. Physical Origin of the Runge-Lenz Vector" *J.Phys.A: Math.* Gen.30, 6831 (1997).

[4] Darwin C.G. Phil. Mag. 39,537 (1920).

[5] Fock V.A. *The Theory of Space, Time and Gravitation.* (Oxford, Pergamon), 1949.

[6] Dirac P.A.M. Forms of Relativistic Dynamics. *Rev. Mod. Phys.* 21,392 (1949).

[7] Argueso F. Sanz J.L. Post-Newtonian Extensions of the Runge-Lenz vector. *J. Math. Phys.* 25, 2936 (1984)

[8] Einstein A., Infeld L., Hoffmann D. Gravitational Equations and Problems of Motion. *Ann. Math.* 39, 65 (1938)

[9] Stahlhofen A.A. Algebraic Solutions of Relativistic Coulomb Problem". Helv. Phys. *Acta*, 70,372 (1997).

Chapter III

DYNAMICAL SYMMETRIES IN NON-RELATIVISTIC QUANTUM MECHANICS

III.1. THE HYDROGEN ATOM (GENERAL CONSIDERATION)

It is natural that the dynamical symmetries appear also in quantum mechanics. It turns out that abovementioned two potentials – Coulomb and oscillator – play the distinguished role even here. Therefore in future our main attention will be carried out on these potentials. As we were convinced of, the additional symmetry in classical mechanics allows us to determine orbits pure algebraically. We'll see that the same happens in quantum mechanics – energy spectrum is obtainable by symmetry considerations only.

Pauli [1] was the first who determined the energy spectrum of the hydrogen atom by using of the algebraic methods of the new matrix mechanics. This verification of the concepts of matrix mechanics was important, but the significance for symmetry technique was not fully appreciated, since the later solution to the same problem using the Schroedinger's wave mechanics was more accessible to physicists, and this latter technique predominated.

Pauli's method was fundamental from another viewpoint. He recognized the role of the LRL observable. He defined LRL vector as follows

$$\vec{A}_0 = \frac{\vec{r}}{r} + \frac{1}{ma}\frac{1}{2}(\vec{L}\times\vec{p} - \vec{p}\times\vec{L}) \tag{III.1}$$

As we see in passing to quantum mechanics we are forced to replace

$$\vec{L}\times\vec{p} \to \frac{1}{2}\left(\vec{L}\times\vec{p} - \vec{p}\times\vec{L}\right)$$

and make transition to the corresponding quantum operators.

The point is that vectors $\vec{L}$ and $\vec{p}$ do not commute any more in quantum mechanics and so product of would be non-commuting quantities has to be defined as a Weyl-ordering product or as a symmetrized product of classical dynamical quantities replaced by the corresponding quantum operators. After such transition is made we are sure that derived quantity is a Hermitian operator and needs not some further manipulations.

At the same time we have an operator of angular momentum $\vec{L} = [\vec{r} \times \vec{p}]$, which obeys the commutation relations (instead of Poisson brackets)

$$[L_i, L_j] = i\varepsilon_{ijk} L_k$$

(we use $\hbar = c = 1$)

Let us now consider the principal aspects of the hydrogen problem.

Algebraic Aspects of the Hydrogen Problem [2]

The Hamiltonian of an electron (charge e, mass m) moving in the Coulomb field of a fixed nucleus (charge Ze) is given by

$$H = \frac{\vec{p}^2}{2m} - \frac{Ze^2}{r} \tag{III.2}$$

It is convenient to use dimensionless variables choosing as the unit of length a/Z, where $a = 1/me^2$ is the radius of the first Bohr orbit. Measuring $\vec{L}$ in units of $\hbar$, momentum $\vec{p}$ in units of $\hbar Z/a$, and energy H in units of $Z^2e^2/2a$, we find for the three operators of interest the dimensionless form

$$\vec{L} = \vec{r} \times \vec{p}$$

$$\vec{A}_0 = \frac{\vec{r}}{r} + \frac{1}{2}\left(\vec{L} \times \vec{p} - \vec{p} \times \vec{L}\right)$$

$$H = p^2 - \frac{2}{r}$$

Notice that the LRL vector $\vec{A}_0$ is a vector operator with respect to $\vec{L}$:

$$[L_i, A_{0j}] = i\varepsilon_{ijk} A_{0k}$$

Square of this operator, $\vec{A}_0^2 \equiv \vec{A}_0 \cdot \vec{A}_0$ is related to the angular momentum operator $\vec{L}^2$ and the Hamiltonian H by the relation

$$\vec{A}_0^2 = H\left(\vec{L}^2 + 1\right) + 1 \tag{III.3}$$

This relation is the operator version of the classical orbit relation.

The properties of the commutator algebra that we obtain for H, $\vec{L}$ and $\vec{A}_0$ are the following (in vectorial notations)

$$\begin{aligned}
&\vec{L}\times\vec{L}=i\vec{L}\\
&\left[L_i, A_{0j}\right]=i\varepsilon_{ijk}A_{0k}\\
&\vec{A}_0\times\vec{A}_0=-i\vec{L}H\\
&\left[H,\vec{L}\right]=\left[H,\vec{A}_0\right]=0
\end{aligned} \qquad \text{(III.4)}$$

One also has

$$\vec{L}\cdot\vec{A}_0=\vec{A}_0\cdot\vec{L}=0$$

There is one significant feature that one should note about this commutator algebra: The Hamiltonian H enters into the commutation relations.

Consider the consequences of this property. Although H is Hermitian, it is not positive definite, and it may, accordingly, posses both positive, zero, and negative eigenvalues of the folowing equation

$$H\psi_E=E\psi_E$$

The structure of the given commutation relations is thus seen to depend on the energy eigenspace on which the operator relations act. Thus, in interpreting these commutation relations in terms of the usual notions of Lie algebras, *we must assume* that the vector space of state vectors on which these equations are allowed to act are energy eigenspaces.

Subject to these conditions, we can now normalize the operator $\vec{A}_0$ as follows:

$$\vec{A}=\begin{cases}\vec{A}_0(-H)^{-1/2}, & for\ \ E<0\\ \vec{A}_0\ , & for\ \ E=0\\ A_0(H)^{-1/2}\ , & for\ \ E>0\end{cases} \qquad \text{(III.5)}$$

Previous commutators can now be put in the Lie algebra form

$$\begin{aligned}
&\vec{L}\times\vec{L}=i\vec{L}\\
&\left[L_i,\ A_j\right]=i\varepsilon_{ijk}A_k\\
&\vec{A}\times\vec{A}=\varepsilon\, i\vec{L}
\end{aligned} \qquad \text{(III.6)}$$

where $\varepsilon = 1, 0, -1$ corresponding to $E < 0,\ E = 0,\ E > 0$, respectively. Under the renorming of $\vec{A}_0$, orbit equations become

$$\begin{aligned}&\left(\vec{L}^2 + \vec{A}^2 + 1\right)H = -1, \qquad &\varepsilon = \pm 1\\ &H\left(\vec{L}^2 + 1\right) + 1 = \vec{A}^2, \qquad &\varepsilon = 0\\ &\vec{L} \cdot \vec{A} = \vec{A} \cdot \vec{L} = 0\end{aligned} \tag{II.7}$$

Let us now discuss the three Lie algebras, corresponding to $\varepsilon = +1, 0, -1$.

For $\varepsilon = +1$ ($E < 0$, spherical geometry) the commutation relations are in the standard $SO(4)$ form of the Lie algebra of the group of 4×4 proper orthogonal matrices $SO(4)$.

For $\varepsilon = -1$ ($E > 0$, hyperbolic geometry) the commutation relations are in the standard $SO(3.1)$ form of the Lie algebra of the Lorentz group.

For $\varepsilon = 0$ ($E = 0$, parabolic geometry) the commutation relations show that the components of LRL vector $\vec{A} = \vec{A}_0$ commute and that $\vec{A}_0$ is a vector operator with respect to $\vec{L}$ (the group is the Euclidean group in three dimensions).

Let us consider the first ($E < 0$) case in more detail [3]. The six generators $\vec{L},\ \vec{A}$ constitute a closed algebra. To clarify this we relabel the indices of the components of $\vec{L}$. First we write $\vec{r} = (r_1, r_2, r_3)$ and $\vec{p} = (p_1, p_2, p_3)$ for which the fundamental commutation relation

$$\left[r_i, p_j\right] = i\delta_{ij}$$

holds and we find, because of

$$\hat{L}_{ij} = r_i p_j - r_j p_i$$

the "natural indices"

$$\hat{\vec{L}} = \left(\hat{L}_{23}, \hat{L}_{31}, \hat{L}_{12}\right)$$

for the $\vec{L}$ – operators. We now extend the indices to $i, j = 1,2,3,4,$ by introducing fourth components r_4 and p_4 that fulfill the canonical commutator and for which

$$A_1 = \hat{L}_{14}, \quad \hat{A}_2 = \hat{L}_{24}, \quad A_3 = \hat{L}_{34}$$

is valid. It is easy to verify that all needed commutation relations are satisfied. The six generators $\hat{L}_{ij}$ obviously constitute a generalization of the three generators $\vec{L}$ from three to four dimensions. The corresponding group can be shown to be the special orthogonal group or the proper rotation group in 4-dimensions, i.e. $SO(4)$. It includes all real orthogonal 4×4 matrices with determinant equal to $+1$. This evidently does not represent a geometrical symmetry of the hydrogen atom since the fourth components r_4 and p_4 are fictitious and can not be identified with geometrical variables. For this reason $SO(4)$ is said to describe a dynamical symmetry of the hydrogen atom. It contains the geometrical symmetry $SO(3)$, generated by the angular momentum operators $\hat{L}_i$, as a subgroup.

It is essential to note that the $SO(4)$ generators are obtained by restriction to *bound states*. Now we derive the energy levels of the hydrogen atom.

It now is comparatively simple to find the energy eigenvalues algebraically. We define the quantities

$$\vec{J}(1)=\frac{1}{2}\left(\vec{L}+\vec{A}_0\right), \qquad \vec{J}(2)=\frac{1}{2}\left(\vec{L}-\vec{A}_0\right)$$

They obey the commutation relations

$$\left[J_i(a),J_j(a)\right]=i\varepsilon_{ijk}J_k(a), \qquad a=1,2$$

$$\left[J_i(1),J_j(2)\right]=0 \tag{III.8}$$

Because of these relations the algebras of $\vec{J}(1)$ and $\vec{J}(2)$ are decoupled; therefore each $\vec{J}(a)$ constitutes $SU(2)$ $(or\ SO(3))$ algebra, and we at once realized that the eigenvalues to be

$$\vec{J}^2(1)=j_1(j_1+1) \text{ and } \vec{J}^2(2)=j_2(j_2+1), \qquad j_i=0,\frac{1}{2},1,\ldots$$

But because of orthogonality

$$(\vec{L}\cdot\vec{A}_0)=\left(\vec{A}_0\cdot\vec{L}\right)=0$$

these invariants (Casimir operators) $J^2(a)$ coincide

$$\vec{J}^2(1)=\vec{J}^2(2)=\frac{1}{4}\left(\vec{L}^2+\vec{A}^2{}_0\right)=j(j+1),$$

where $j \equiv j_1 \equiv j_2$.

Then according to earlier relation (III.7), we obtain for the energy spectrum

$$E = -\frac{1}{2(2j+1)^2}, \qquad j = 0, \frac{1}{2}, 1, \ldots$$

which is nothing but the famous Balmer formula, where $2j+1$ plays the role of a principal quantum member

$n = 2j + 1 = 1,\ 2, \ldots$

Also the degeneracy of states is reproduced correctly: $\vec{J}(1)$ and $\vec{J}(2)$ can each have $2j+1$ independent eigenvalues, and therefore there are $(2j+1)^2$ states. Also, we recognized that $\vec{L}$ is an axial vector, not changing its sign on space inversion. But $\vec{A}$ is a polar vector, which does change sign. We thus expect that states characterized by the symmetry generators $\vec{L}$ and $\vec{A}$ need not have well-defined parity. This is actually the case, since states of even and odd l are degenerate in the hydrogen atom.

Therefore we saw that to the level of hydrogen atom with fixed n corresponds irreducible representation $D\left(\frac{n-1}{2}, \frac{n-1}{2}\right)$ of the group $SO(4)$. Dimension of this representation (i.e. a multiplicity of degeneracy) is equal to $(2j+1)^2 = n^2$.

III.2. The Hydrogen Atom in the Momentum Representation

After making a Fourier transformation the Schroedinger equation for Coulomb potential is written in the momentum representation as follows

$$\left(\frac{\vec{p}^{\,2}}{2} - E\right)\phi(\vec{p}) = \frac{1}{2\pi^2}\int\frac{d^3q\,\phi(\vec{q})}{|\vec{q} - \vec{p}|^2} \tag{III.9}$$

It is convenient to consider momentum space as a stereographic projection of 4-dimensional hypersphere. Following V. Fock [4], let us introduce coordinates ξ_μ,

$$\vec{\xi} = \frac{2p_0\vec{p}}{p_0^2 + p^2}; \quad \xi_4 = \frac{p_0^2 - p^2}{p_0^2 + p^2}, \qquad \left(p_0^2 = -2mE > 0\right)$$

They obey

$$\xi^{\mu}\xi_{\mu} = \vec{\xi}^{\,2} + \xi_4^2 = 1$$

Let us now introduce a new function $\psi(\xi)$, defined on unit hypersphere

$$\psi(\xi) = const\left(p_0^2 + p^2\right)^2 \phi(\vec{p})$$

and taking into account the following relations

$$(\xi - \xi')^2 = \frac{4p_0^2\left|\vec{p} - \vec{p}'\right|^2}{\left(p_0^2 + p^2\right)\left(p_0'^2 + p'^2\right)}, \quad d\vec{p} = \left(\frac{p_0^2 + p^2}{2p_0}\right)^2 \frac{d\xi}{\xi_4}$$

our equation (III.9) reduces to the following form [5]

$$\psi(\xi) = \frac{1}{2\pi^2 p_0} \int \frac{d\xi'}{\xi_4'} \frac{\psi(\xi')}{(\xi - \xi')^2} \quad \text{(III.10)}$$

Integration in this equation is taken on the whole surface of the 4-dimensional hypersphere. $\frac{d\xi}{\xi_4}$ is the invariant volume element on this surface. Indeed, introducing as usually spherical coordinates (α, θ, φ) in this hypersphere

$$\xi_1 = \sin\alpha \sin\theta \cos\varphi$$
$$\xi_2 = \sin\alpha \sin\theta \sin\varphi$$
$$\xi_3 = \sin\alpha \cos\theta$$
$$\xi_4 = \cos\alpha$$
$$(0 \le \alpha,\ \theta \le \pi; \quad 0 \le \varphi \le 2\pi)$$

we derive

$$\frac{d\xi}{\xi_4} = \sin^2\alpha \sin\theta d\alpha d\theta d\varphi = d\Omega$$

It is evident that $d\Omega$ is not changed in the case of 4-dimensional rotations. Therefore the obtained equation (III.10) is invariant under the transformations of the $SO(4)$ group. But as it is nothing more than the Schroedinger equation for the hydrogen atom in ξ-representation, then it is established that the "hidden symmetry" group for the Coulomb problem in quantum mechanics is $SO(4)$ in case of $E < 0$.

We remark here that it is easy task to generalize this result to arbitrary D-dimensions [6] – it is sufficient to consider the relevant stereographic projection on the unit hypersphere in $(D+1)$-dimensions and perform suitable geometric transformations. In any case we derive symmetry group $SO(D+1)$, larger than the group of ordinary $SO(D)$ rotations.

Example of Application of The Momentum Representation: Dynamical Symmetry of a Three Dimensional Wick-Cutkosky Problem [7]

It is well known that the bound state problem of two scalar particles interacting via the scalar "photons" is exactly solved in the framework of the Relativistic Bethe-Salpeter equation in the ladder approximation. This problem has exactly been solved by Wick [8], and it turn out that the solution of the zero energy equation possesses the higher dynamical $O(5)$ symmetry. Cutkosky [9] considered the case of non-zero angular momentum and obtained the complete set of solutions.

On the other hand, the comparision of spectra of the quasipotential equation of Logunov ad Tavkhelidze [10] with the Bethe-Salpeter ones shows that the first gives a good description of relativistic effects. At the same time, it does not suffer from shortcomings characteristic of the Bethe-Salpeter equation: the existence of the "relative" time, the absence of probability interpretation and so on.

Hence it is interesting to look for solvable models in a three-dimensional quantum field theory and to study higher symmetry properties of solutions corresponding to quasipotential equations.

Let first study the dynamical symmetry of the Wick problem:

The Bethe-Salpeter equation for the equal mass scalar particles exchanging massless and scalar photon in the centre of mass system with zero total energy has the form

$$\left(p^2+m^2\right)^2\Phi(p)=\frac{\lambda}{\pi^2}\int\frac{\Phi(k)d^4k}{(p-k)^2} \tag{III.11}$$

where p is a relative 4-momentum in the CM system and Wick's rotation [8] is performed.

After mapping a four-dimensional momentum space upon the surface of a five-dimensional sphere by a stereographic projection, we derive the $O(5)$ invariant equation for a five-dimensional harmonics [4]

$$H(\alpha)=\frac{\lambda}{16\pi^2}\int\frac{d\Omega'_S H(\alpha')}{\sin^2\frac{\theta}{2}} \tag{III.12}$$

where $H(\alpha) = \sec^6 \frac{\alpha}{2} \Phi(\alpha)$

Equation (III.12) is invariant under all the rotations of 5-sphere, $\lambda = \lambda_N = N(N+1)$ and the solution may be presented as a five-dimensional spherical harmonics

$$\Phi_{nl}^{m} = \int_{-1}^{1} g_n(z) dz \frac{Y_l^m(\vec{p}) C_{n-l-1}^{l+1}\left(\frac{X}{R}\right) R^{n-l-1}}{\left(m^2 + p^2\right)^{2+n}} \qquad \text{(III.13)}$$

where

$$R^2 = \left(p^2 + m^2\right)^2 + 4p_0^2, \qquad X = m^2 - p^2$$

The functions $g_n(z)$ of the form [5]

$$g_n(z) = \left(1 - z^2\right)^n C_k^{n+1/2}(z)$$

satisfy the differential equation of the Gegenbauer polynomials and have the following eigenvalues $\lambda = (n+k)(n+k+1)$, where k is a new quantum number which is set to zero for the right correspondence with non-relativisic case.

Now we return to the quasipotential equation for the Wick problem. This equation has the form [10]:

$$\left[E^2 - 4\left(\vec{p}^{\,2} + m^2\right)\right]\Psi(\vec{p}) = \int \frac{d^3 k}{\sqrt{\vec{p}^{\,2} + m^2}} V\left(\vec{p} - \vec{k}\right)\Psi\left(\vec{k}\right) \qquad \text{(III.14)}$$

where $\vec{p}$ and $\vec{k}$ are three-dimensional relative momenta and $\Psi(\vec{p})$ is the quasipotential wave function of the bound state with the following normalization condition:

$$\int \Psi^*(\vec{p}) \widetilde{F}_0^{-1} \Psi(\vec{p}) d^3 p = 1 \qquad \text{(III.15)}$$

It is an eigenfunction of the coupling constant problem at fixed energy. $\widetilde{F}_0$ is the operator of the free Green function of the quasipotential equation.

There is connection between the quasipotential and the Bethe-Salpeter wave functions [10]

$$\Psi(\vec{p}) = \int_{-\infty}^{+\infty} dp_0 \Phi(\vec{p}, p_0) \qquad \text{(III.16)}$$

Let us apply the quasipotential equation (III.14) to the Wick-Cutkosky problem. The choice of suitable quasipotential is important. The quasipotential may be derived for the relativistic interaction by the prescription given in [10]. It is also possible to restore potential from quantum field theory by the corresponding Feynman diagrams.

Since the Wick model is a problem of searching for the zero energy bound states if the Coulomb interaction exists, we consider the standard procedure allowing derivation of the potential from quantum field theory.

We should like to remark that an important feature of the Coulomb potential in the momentum space is the dependence of the description power on the dimension of the space

$$\frac{1}{r}=\left(\pi\omega_{N-1}\right)^{-1}\int\frac{d^N q e^{i\vec{q}\vec{r}}}{|\vec{q}|^{N-1}}, \qquad \omega_N=\frac{2\pi^{-N/2}}{\Gamma(N/2)}$$

The kernel of the potential $V(\vec{p}-\vec{k})=\left|\vec{p}-\vec{k}\right|^{N-1}$ in an $(N+1)$-dimensioal space is the Green function of the Laplace operator

$$\Delta_u^{N+1}\left(\left|\vec{u}-\vec{\upsilon}\right|^{N-1}\right)^{-1}=-(N+1)\omega_{N+1}\delta^{N+1}(\vec{u}-\vec{\upsilon})$$

It turned out that the quasipotential for spinless, equal mass particles, interacting through a massless scalar particle, evaluated from the second order perturbation theory in the framework of quantum electrodynamics has the following form in the zero energy limit [12]:

$$V\left(\vec{p}-\vec{k}\right)_{E\to 0}=\frac{1}{\sqrt{\vec{p}^2+m^2}}\frac{1}{\left|\vec{p}-\vec{k}\right|} \tag{III.17}$$

If we take this into account then (III.14) becomes analogous to a three-dimensional Bethe-Salpeter equation of the Wick problem, and we shall show below that it describes Wick model in terms of quasipotential

$$\left(\vec{p}^2+m^2\right)^2=\frac{\lambda}{4\pi}\int\frac{\Psi\left(\vec{k}\right)d^3k}{\left|\vec{p}-\vec{k}\right|} \tag{III.18}$$

For the case of zero angular momentum equation (III.18) is equivalent to the following differential equation:

$$\frac{1}{p}\frac{d}{dp^2}\left[p\left(p^2+m^2\right)^2\Psi(p)\right]+\lambda\Psi(p)=0 \tag{III.19}$$

with the boundary conditions

$$p\Psi(p) \to 0, \qquad p \to 0$$
$$p\left(p^2 + m^2\right)\Psi(p) \to const, \qquad p \to \infty$$

The ground state solution of the s-wave problem $\Psi_{n=1}(\vec{p}) = C\left(\vec{p}^{\,2} + m^2\right)^{-5/2}$ coincides with the corresponding integrated solution of the Bethe-Salpeter equation (III.13).

In the configuration space it takes the form

$$\Psi_{n=1}(r) = \frac{4\pi}{3}\left(\frac{r}{m}\right)K_1(mr)$$

Owing to the well-known features of the modified first kind Bessel function $K_1(z)$, this solution has correct behaviour at zero and characteristic asymptotics for the bound state wave function.

If we consider the case of non-zero angular momentum, we deal with the equation

$$\nabla_{\vec{p}}^2\left[\left(\vec{p}^{\,2} + m^2\right)^2\Psi(\vec{p})\right] + \lambda\Psi(\vec{p}) = 0$$

where the Laplace operator contains the angular variables

$$\nabla_{\vec{p}}^2 = \nabla_p^2 + \frac{1}{p^2}\nabla_{\theta,\varphi}^2$$

If we suppose that a solution of the equation has the form $\Psi(\vec{p}) = R_{nl}(p)Y_{lm}(\theta,\varphi)$, we get the following radial equation:

$$\frac{1}{p^2}\frac{d}{dp}\left[p^2\frac{d}{dp}\left(p^2 + m^2\right)^2 R_{nl}(p)\right] + \left[\lambda - \frac{l(l+1)}{p^2}\left(p^2 + m^2\right)^2\right]R_{nl}(p) = 0$$

which after substitution of the variables $p^2 = u$, $\quad R = u^{-1/4}\left(u + m^2\right)^{-2}\Phi(u)$ takes the form

$$\left[\frac{d^2}{du^2} + \frac{1}{u}\frac{d}{du} + \frac{\lambda}{4u\left(u + m^2\right)^2} + \frac{1/4 - l(l+1)}{4u^2}\right]\Phi_{nl}(u) = 0 \qquad \text{(III.20)}$$

which is a differential equation of the adjoint spherical functions $P_\mu^\nu(z)$ [12].

Therefore obtained solutions are

$$R_{nl}(p)=\frac{1}{\sqrt{p}\left(p^2+m^2\right)^{\frac{1}{2}}}P_{n-1/2}^{-(l+1/2)}\left(\frac{m^2-p^2}{m^2+p^2}\right)$$

and $\lambda = 4n^2 - 1,\qquad n = 1, 2,\ldots$ (III.21)

Now we show that the Fock stereographic projection method [4] applied with the quasipotential equation (III.18) gives the solution of (III.20).

For the purpose of confirming the above-mentioned, let us introduce the stereographic coordinates on the surface of a unit 4-sphere like as in previous Section:

$$\vec{\xi}=\frac{2m\vec{p}}{m^2+\vec{p}^{\,2}},\qquad \xi_4=\frac{m^2-\vec{p}^{\,2}}{m^2+\vec{p}^{\,2}},\qquad \xi^2=\vec{\xi}^{\,2}+\xi_4^2=1 \qquad \text{(III.22)}$$

or

$$\xi_1=\sin\alpha\sin\theta\sin\varphi \qquad \frac{|\vec{p}|}{m}=\tan\frac{\alpha}{2}$$

$$\xi_2=\sin\alpha\sin\theta\cos\varphi \qquad \frac{|\vec{p}|}{\sqrt{m^2+\vec{p}^{\,2}}}=\sin\frac{\alpha}{2}$$

$$\xi_3=\sin\alpha\cos\theta$$

$$\xi_4=\cos\alpha \qquad \frac{m}{\sqrt{m^2+\vec{p}^{\,2}}}=\cos\frac{\alpha}{2}$$

The connection between the volume elements in the new coordinates and in the four-dimensional Euclidean space is given by

$$d^3k=\left(\frac{m}{2}\sec^2\frac{\alpha}{2}\right)^3 d^4\Omega$$

and equation (III.18) takes the form

$$H(\alpha)=\lambda_n\int\frac{H(\alpha')d^4\Omega'}{\sin\chi/2},\qquad \lambda_n=\frac{\sqrt{2}}{32\pi}\lambda \qquad \text{(III.23)}$$

where $H(\alpha)=\cos^5\frac{\alpha}{2}\Psi(\alpha)$ and

$$\cos\chi = \cos\alpha\cos\alpha' + \sin\alpha\sin\alpha'\cos\gamma$$
$$\cos\gamma = \cos\theta\cos\theta' + \sin\theta\sin\theta'\cos(\varphi - \varphi')$$

is cosine of an angle between the vectors on the surface of a 4-sphere.

Equation (III.13) is now written in an obviously $O(4)$ -symmetric form.

If we exploit corresponding mathematical relations [12],we may derive the eigenfunctions

$$H(\alpha,\theta,\varphi) = P_{nl}(\alpha)Y_{lm}(\theta,\varphi)$$
$$P_{nl}(\alpha) = \frac{1}{\sqrt{|\sin\alpha|}} P_{n-1/2}^{-(l+1/2)}(\cos\alpha), \qquad n = 1, 2, \ldots$$

and eigenvalues

$$\lambda_n^{-1} = \frac{4\pi}{n+1}\int_{-1}^{1}\frac{dx}{\sqrt{1-x}} C_n^1(x)\left(1-x^2\right)^{1/2} = \frac{16\pi\sqrt{2}}{4(n+1)^2 - 1}.$$

Thus the solution of equation (III.8) takes the form

$$\Psi_{nl}(\vec{p}) = \frac{1}{\left(m^2+p^2\right)^2}\frac{1}{\sqrt{2p}} P_{n-1/2}^{-(l+1/2)}\left(\frac{m^2-p^2}{m^2+p^2}\right) Y_{lm}(\theta,\varphi)$$

This is in complete accordance with the previous result.

III. 3. The Hydrogen Atom and the Lorentz Group

We were convinced above that the states of the Hydrogen atom with discrete spectrum of energy are described by irreducible representations of $SO(4)$ group. The following question arises: is it possible the unified description of both discrete and continuous spectra in the language of the group theory? This is not achievable in the framework of $SO(4)$, because all its representations are finite–dimensional (as for all compact groups, i.e. for continuous group with finite volume).

Therefore the "full" group of dynamical symmetry must be non–compact. It seems that this is a Lorentz group.

In order to see this, let us begin with $E > 0$ states, and introduce an operator

$$\vec{A}' = \frac{1}{\sqrt{2H}}\vec{A}$$

Then the commutation relations take the form

$$\begin{aligned} &[L_i, L_j] = i\varepsilon_{ijk} \\ &[L_i, L_j] = i\varepsilon_{ijk} A'_k \\ &[A'_i, A'_j] = -i\varepsilon_{ijk} L_k \end{aligned} \qquad \text{(III.24)}$$

which coincides with the commutation relations for generators of the homogeneous Lorentz group.

The Lorentz group has two kinds of irreducible representations [5]:

Finite–dimensional representations $D(j_1, j_2)$, which are non–unitary representations, except $D(0,0)$.

Infinite–dimensional unitary representations, opened by Gel'fand and Naymark [13] .

These last ones, by themselves, are divided on two classes (representations of principal and additional series, from which only principal class is used in physics. It is denoted by $D(m,\rho)$. Here m is integer, but ρ - arbitrary real number. $D(m,\rho)$ and $D(-m,-\rho)$ are equivalent, therefore $\rho > 0$ is considered as a rule (for any signs of m). Invariant operators F and G of the Lorentz group take the following values on $D(m,\rho)$:

$$F = L^2 - A'^2 = -\left[1 + \frac{1}{4}\left(\rho^2 - m^2\right)\right]$$

$$G = \left(\vec{L} \cdot \vec{A}'\right) = \frac{m\rho}{4}$$

By analytic continuation it is possible to derive usual finite–dimensional representations, considering relations

$$m = 2(j_1 - j_2), \qquad \rho = -2i(j_1 + j_2 + 1)$$

For finite dimensional representations it follows that $\rho = -ik \quad (k = 2,3,4,...)$. It can be shown, that exact representations of the Lorentz group correspond only to points $0 \le \rho < +\infty$ and integer points $\rho = -ik$ lying on the imaginary axis. Hamiltonian H is expressed in the following way

$$H = -\frac{1}{2}(F+1)^{-1} = -\frac{1}{2}\left(\vec{L}^2 - \vec{A}'^2 + 1\right)^{-1}$$

It follows then that

$$F=-\left(1+\frac{1}{2E}\right),\qquad G=-\frac{1}{\sqrt{2H}}\left(\vec{L}\cdot\vec{A}\right)=0$$

Eigenvalues of F and G (Casimir's operators) determine uniquely the irreducible representation:

$$\rho=\begin{cases}\sqrt{\dfrac{2}{E}}, & \textit{for continuous spectrum}\\ -2in, & \textit{for discrete spectrum}\end{cases}$$

System of continuous spectrum wave functions with a given E forms infinite-dimensional unitary representation $D(0,\rho)$ of the Lorentz group. Physically infinite–dimensionality means that values of orbital momentum l in the continuous spectrum are not bounded; Unitarity follows from the fact, that generators L_i and A'_j are hermitian for $H>0$. States of discrete spectrum, corresponding to the level with a principal quantum number n, form a finite–dimensional representation $D(j_1,j_2)$ with $j_1=j_2=\dfrac{n-1}{2}$. This representation is not unitary, because A'_i generators becomes anti-hermitian for $E<0$. The multiplicity of this representation is equal to $(2j_1+1)(2j_2+1)=n^2$. When n changes from 1 to ∞, j takes all integer and half – integer values: $j=0,\frac{1}{2},1,\frac{3}{2},\ldots,$

Therefore all representations of the Lorentz group with $m=0$ are realized on the hydrogen atom's wave functions. This restriction is natural, because nonzero values of m appear only in case when particles have spin.

It is known, that there is a one to one correspondence between the finite – dimensional representations of $O(4)$ group and those of the Lorentz group: by analytic continuation of $D(j_1,j_2)$ to the range of imaginary angles of rotation in (x_i,x_0)-plane the corresponding Lorentz representation creates. Therefore if one is restricted only by consideration only bound states than the group $O(4)$ can be considered as a group of a hidden symmetry. But in this case we lose the continuous spectra, for a description of which it is necessary to pass to non – compact group. This unables us to describe uniquely both discrete and continuous spectra. Therefore it is natural to consider just Lorentz group as a group of dynamical symmetry of the hydrogen atom.

III. 4. Three Dimensional Isotropic Harmonic Oscillator and Su(3) [14]

The Hamiltonian of the harmonic oscillator is

$$H = \frac{1}{2}\left(\vec{p}\cdot\vec{p} + \omega^2 \vec{r}\cdot\vec{r}\right)$$

where we have chosen units so that the mass equals unity. In coordinate representation where $\vec{p} = -i\vec{\nabla}$, the solution of the eigenvalue problem becomes a matter of seeking the solution of a second–order partial differential equation. Usually the method of separation of variables is employed. The separation constants that appear in this method have only discrete values for the bound–state problem, and moreover represent some physical attribute of the system.

For the harmonic oscillator, the eigenvalue problem separates in spherical coordinates, and the operators to the separation constants are just the components of the familiar angular momentum vector–operator

$$\vec{L} = \vec{r} \times \vec{p}$$

In addition, the eigenvalue problem also separates in Cartesian coordinates and by inspection we know that the operators corresponding to the three separation constants are $\frac{1}{2}\left(p_i p_j + \omega^2 r_i r_j\right)$, no sum on i. This has the appearance of diagonal components of a symmetric tensor operator

$$A_{ij} = \frac{1}{2}\left(p_i p_j + \omega^2 r_i r_j\right) \qquad \text{(III.25)}$$

as expected

$$\left[A_{ij}, H\right] = 0$$

In addition, this operator has the following algebraic properties

$$\sum_j A_{ij} L_j = \sum_i L_i A_{ij} = 0 \qquad \text{(III.26)}$$

$$\sum_j A_{ij} A_{jk} = HA_{ik} + \frac{1}{2}\omega^2\left(L_i L_k - \delta_{ik} L^2 + 2\left[L_i, L_k\right] - 2\delta_{ik}\right) \qquad \text{(III.27)}$$

$$\text{Trace } A = \sum_i A_{ii} = H$$

$$\sum_{i,j} r_i \left(H\delta_{ij} - A_{ij}\right) r_j = \frac{1}{2} L^2$$

$$\sum_{i,j} p_i \left(H\delta_{ij} - A_{ij}\right) p_j = \frac{1}{2}\omega^2 L^2$$

$$\sum_{i,j} p_i \left(H\delta_{ij} - A_{ij}\right) r_j = \sum_{i,j} r_i \left(H\delta_{ij} - A_{ij}\right) p_j = 0$$

$$A_{ii}A_{jj} - \left(A_{ij}\right)^2 = \frac{1}{4}\omega^2 \sum_k \left(\varepsilon_{ijk}\right)^2 \left(L_k^2 + 1\right) \text{ - no sum on } i \text{ or } j$$

We see that the tensorial trace of A_{ij} is just the Hamiltonian. The remaining five independent operators of the traceless symmetric tensor are conveniently given by the spherical components

$$\begin{aligned} A_\theta &= \omega^{-1}\left(2A_{33} - A_{11} - A_{22}\right), \\ A_\epsilon &= -\varepsilon\omega^{-1}\left(A_{13} + i\varepsilon A_{23}\right), \qquad \varepsilon = \pm 1 \\ A_{2\varepsilon} &= \omega^{-1}\left(A_{11} - A_{22} + 2i\varepsilon A_{12}\right), \qquad \varepsilon = \pm 1 \end{aligned} \tag{III.28}$$

all of which commute with the Hamiltonian. These operators and the angular momentum operators, written in spherical component form as

$$L_3 \text{ and } L_\varepsilon = L_1 + i\varepsilon L_2, \quad \varepsilon = \pm 1$$

are closed under the process of commutation.

Specifically, the commutation relations are such [14], that the eight operators comprising the five operators derived form the traceless symmetric tensor operator and the three components, characteristic of the infinitesimal operators of the $SU(3)$ group. These operators may be brought into correspondence with the standard form for the operators of the $SU(3)$ group, having a symmetrical root diagram.

A more direct approach is to consider the Hamiltonian written terms of destruction and creation operators. Then

$$H = \frac{\omega}{2}\sum_{i=1}\left(b_i^+ b_i + b_i b_i^+\right) = \frac{3\omega}{2} + \omega\sum_{i=1}^{3} b_i^+ b_i$$

where

$$b_i = \frac{1}{\sqrt{2\omega}}\left(\omega r_i + ip_i\right)$$

$$b_i^+ = \frac{1}{\sqrt{2\omega}}\left(\omega r_i - ip_i\right)$$

These boson operators have the usual commutation relations

$$\left[b_i, b_j^+\right] = +\delta_{ij}; \quad \left[b_i, b_j\right] = \left[b_i^+, b_j^+\right] = 0$$

It is apparent that the transformation

$$b_i \to b_i' = \sum_j U_{ij} b_j$$

leaves the Hamiltonian and the commutation laws unchanged if $U^+ = U^{-1}$. Hence, the invariance under $SU(3)$ is immediately established.

A tensor basis of this algebra is provided by the form $b_i^+ b_j - \frac{1}{3}\delta_{ij}\sum_k b_k^+ b_k$. This leads to an eight-dimensional representation, which for $SU(3)$ is the regular representation.

Since the infinitesimal operators themselves may be taken as the basis of the regular representation, this suggests that the operators may be formed from the combinations of $b_i^+ b_j$. In fact if we define the tensor

$$B_{ij} = b_i^+ b_j + b_j b_i^+$$

then

$$\left[B_{ij}, H\right] = 0$$

$$A_{ij} = \frac{\omega}{4}\left(B_{ij} + B_{ji}\right)$$

$$L_i = -\frac{i}{2}\varepsilon_{ijk} B_{jk}$$

In terms of the destruction and creation operators the standard elements of the Lie algebra for $SU(3)$ are

$$A_0 = 2b_3^+ b_3 - \left(b_+^+ b_+ + b_-^+ b_-\right)$$

$$L_3 = -\left(b_+^+ b_+ - b_-^+ b_-\right)$$

$$\hat{E}_3^\lambda = \varepsilon \frac{1}{\sqrt{6}}\left(\delta_{-1,\varepsilon\lambda} b_3^+ b_3 - \delta_{1,\varepsilon\lambda} b_{-\varepsilon} b_3^+\right)$$

$$\hat{E}_{2\varepsilon} = \frac{1}{\sqrt{6}} b_{-\varepsilon}^+ b_\varepsilon$$

Here the spherical component destruction and creation operators, defined by

$$b_\varepsilon = \frac{1}{\sqrt{2}}(b_1 + i\varepsilon b_2), \qquad b_\varepsilon^+ = \frac{1}{\sqrt{2}}(b_1^+ - i\varepsilon b_2^+)$$

have corresponding commutation relations

$$[b_\varepsilon, b_{\varepsilon'}^+] = \delta_{\varepsilon\varepsilon'}; \qquad [b_\varepsilon, b_\varepsilon] = [b_\varepsilon^+, b_{\varepsilon'}^+] = 0$$

REFERENCES

[1] Pauli W, Uber dasWasserstoffspectrumom Standpunkt dr neuen Quanten- mechanik, *Z.Phys*. 36, 336 (1926).

[2] Biedenharn L.C., Louk L.D., Angular Momentum in Quantum Physics, in "Encyclopedia of Mathematics and its Application", Addison-Wisley Pub. Company, p.335 (1981).

[3] Greiner W. Quantum Mechanics. Symmetries., *Springer,* p.453 (1994).

[4] Fock V.A., Zur Theorie des Wasserstoffatoms, *Z.Phys*. 145 (1935).

[5] Popov V.S. On Hidden Symmetry of the Hydrogen Atom, In Proceedings "Physics of High Energy and Elementary Particles". *Naukova dumka*, Kiev, 1967. p.702.

[6] Bander M. and Itzykson C. Group Theory and the Hydrogen Atom, (I) and (II). *Rev.Mod.Phys*.18,330;346 (1966).

[7] Matveev V.A., Slepchenko L.A. and Vardiashvili M.D. Dynamical Symmetry of a Three-Dimensional Wick-Cutkosky Problem. Preprint, E2-88-695 (1988)

[8] Wick G. Properties of Bethe-Salpeter Wave Functions. Phys.Rev. 96, 1124 (1954).Cutkosky R. Solution of a Bethe-Salpeter Equation. *Phys.Rev*. 96,1135 (1954).

[9] Logunov A.A. and Tavkhelidze A.N. Quasi-otpical Approach in Quantum Field *Theory.Nouvo Cim*. 29, 380 (1963).

[10] Nguyen Van Hieu ad Faustov R.N. Quasipotential in Quantum Electrodynamics. *Nucl.Phys*.53, 337 (1964).

[11] Gradshtein I.S. and Ryzhik I.M. Tables of Integrals, Series and Products. NY,Academic, 1965.

[12] Gel'fand I.M., Graev M.I. Irreducible Representations of Lie Algebra of U(p,q) Group. In Proceedings "Physics of High Energy and Elementary Particles", *Naukova Dumka,* Kiev, 1967.p. 216.

[13] Naymark M.A. Unitary Representations of non-Compact Gpoups. In Proc. "Physics of High Energy and Elementary Particles", *Naukova Dumka,* Kiev,1967, p. 183.

[14] Fradkin D.M., Three Dimensional Isotropic Harmonic Oscillator and SU(3), *Am.J.Phys*. 33, 207 (1965).

Chapter IV

A New Kind of Dynamical Symmetry – Supersymmetry

"...there are symmetries in the world
which are difficult to discover
because they are spontaneously broken."

Lewis H. Ryder

IV.1 Supersymmetric Quantum Mechanics

Physicists have long strived to obtain a unified description of all basic interactions in nature, i.e., strong, electroweak and gravitation interactions. Several ambitious attempts have been made in the last three or four decades, and it is now widely felt that supersymmetry (SUSY) is a necessary ingredient in any unified approach [1].

What is remarkable about SUSY is that it "mixes" fermions and bosons. If SUSY where exact there would exist in the world fermions and bosons of the same masses. We know that in nature this is simply not the case: Despite the beauty of all known unified theories, there has so far been no experimental evidence of SUSY being realized in nature.

SUSY is the relativistic symmetry between fermions and bosons. It is the only known way to unify space–time and internal symmetries in the relativistic S – matrix theory.

The main symmetry in relativistic quantum theory is the Poincare group, which contains Lorentz transformations and space–time translations. The fundamental significance has also the symmetry between identical particles, which gives the division of particles into fermions and bosons. All the traditional transformations do not mix particles with the different spins.

It might appear as paradoxical at first glance that SUSY connects particles with integer and half–integer spins, while they obey the different statistics.

Why is so much interests about SUSY despite the fact that there is no trace of this symmetry and its features are not manifested anywhere? SUSY continues to attract the mathematicians and physicists alike who are asked to come to terms with new ideas and concepts which the theory exposes from time to time.

The most essential reasons are:

- The possibility of construction of non–contradictory theory for such a symmetry;
- In case of ordinary symmetry (Lie) groups some kinds of "no-go" theorems forbid nontrivial unification of space-time and internal symmetries. In SUSY theory these theorems do not work, because of the grassmanian nature of relevant transformation parameters;
- In SUSY field theory, owing to the strong restrictions cancellation of divergences occur, and as a result, some SUSY models become not only renormalizable, but even finite [1].

It was in the context of these properties that SUSY was first studied in the simplest case of SUSY quantum mechanics[2-4], as a testing ground for the non – perturbative methods of seeing SUSY breaking in field theory.

There have also been applications of SUSY in atomic, condensed matter and statistical physics.

In the discussion of SUSY quantum mechanics one usually considers one– dimensional systems.

First of all, after the introduction of the basic state vectors

$$|n_B, n_F\rangle, \quad n_B = 0,1,\ldots,\infty, \quad n_F = 0,1$$

where n_B and n_F are the filling numbers for bosons and fermions, respectively, one needs to determine the operators $Q_{\pm}$, that convert bosons into fermions and vice versa. In the simplest case these operators act in the following manner

$$Q_+|n_B, n_F\rangle \sim |n_B - 1,\, n_F + 1\rangle, \quad if \;\; n_F = 0$$
$$Q_-|n_B, n_F\rangle \sim |n_B + 1,\, n_F - 1\rangle, \quad if \;\; n_F = 1$$

They are called as fermionic generators and are expressed by means of simple destruction and creation operators as

$$Q_+ = qbf^+, \quad Q_- = qb^+ f \tag{IV.1}$$

where q is an arbitrary real constant, the same for both $Q_{\pm}$, in order $Q_{\pm}^+ = Q_{\mp}$. The $b(bosonic)$ and $f\,(fermionic)$ operators obey

$$\begin{aligned} &[b, b^+] = 1, \\ &\{f, f^+\} \equiv ff^+ + f^+ f = 1, \\ &f^+ f^+ = ff = 0, \\ &[b^i, f^i] = 0 \end{aligned} \tag{IV.2}$$

This property of fermionic operators $f^2 = 0$, is called a nilpotency. Because of this property $Q_\pm$ are also nilpotents

$$Q_+^2 = Q_-^2 = 0$$

One can introduce two Hermitian operators

$$Q_1 = Q_+ + Q_-$$
$$Q_2 = -i(Q_+ - Q_-)$$

It is evident that they are anti–commuting:

$$\{Q_1, Q_2\} = 0$$

moreover their squares are the same

$$Q_1^2 = Q_2^2 = \{Q_+, Q_-\}$$

These relations dictate the simplest form of Hamiltonian $\widetilde{H}$, which could be invariant under the supersymmetric transformations

$$\widetilde{H} = Q_1^2 = Q_2^2 = \{Q_+, Q_-\}$$

From the nilpotency of $Q_\pm$ operators it follows the supersymmetry of this Hamiltonian

$$[H, Q_i] = 0 \quad (i = \pm)$$

The fact that supercharges $Q_\pm$ commute with H is responsible for the degeneracy. These operators can be interpreted as operators which change bosonic degrees of freedom into femionic ones and vice versa.

Above derived relations can be rewritten in the unified form

$$\{Q_i, Q_j\} = 2\delta_{ij}\widetilde{H},$$
$$\left[Q_i, \widetilde{H}\right] = 0 \qquad \text{(IV.3)}$$

It is the simplest Lie superalgebra. Often $\widetilde{H}$ is called as Witten's Hamiltonian, and the algebra – as Witten's algebra [5]. This algebra contains both commutators and anticommutators (graded algebra) and characterizes a new kind of symmetry –

supersymmetry. The dynamical feature of supersymmetry is manifested in that the Hamiltonian $\widetilde{H}$ enters among the generators of supersymmetry.

As a rule Witten's Hamiltonian is expressible by the original Hamiltonian of quantum system $\widetilde{H} = f(H)$, where $f(H)$ is invertable function of H.

Let us list the general features of SUSY quantum mechanics:

- The spectrum of $\widetilde{H}$ is positive semi–definite $(E \geq 0)$.
- All levels are two-fold degenerates, except $E = 0$ (the ground state).
- Let express $\widetilde{H}$ in terms of operators (b, f):

$$\begin{aligned}\widetilde{H} &= \{Q_+, Q_-\} = q^2\left(b^+b + f^+f\right) = \\ &= q^2\left(b^+b + \tfrac{1}{2}\right) + q^2\left(f^+f - \tfrac{1}{2}\right) = \\ &= \widetilde{H}_B + \widetilde{H}_F\end{aligned} \tag{IV.4}$$

This result tells us that $\widetilde{H}$ is the sum of non–interacting bosonic $\widetilde{H}_B$ and fermionic $\widetilde{H}_F$ oscillators, with energies

$$\widetilde{H}_B = q^2\left(b^+b + \tfrac{1}{2}\right), \quad E_B = q^2\left(n_B + \tfrac{1}{2}\right), \quad n_B = 0,1,\ldots,\infty$$
$$\widetilde{H}_f = q^2\left(f^+f - \tfrac{1}{2}\right), \quad E_F = q^2\left(n_F - \tfrac{1}{2}\right), \quad n_F = 0,1$$

They have the equal frequencies $\omega^2 = q^2$. This provides the supersymmetry of $\widetilde{H}$.

The total energy of SUSY oscillator is

$$E_{n_b, n_F} = \omega\left(n_B + n_F\right)$$

Let us mention that the ground state (vacuum) energy is zero. The zero frequency energy of the bosonic oscillator (which is infinite in quantum field theory) is compensated by the negative zero–frequency energy of the fermionic oscillator. It is the magic property of supersymmetry. All the levels of $\widetilde{H}$ oscillator are degenerate two-fold, except the ground state level.

The interaction in SUSY quantum mechanics are introduced by generalization of simple bosonic operators:

$$\begin{aligned}Q_+ &= B\left(b, b^+\right)f^+ \\ Q_- &= B^+\left(b, b^+\right)f\end{aligned} \tag{IV.5}$$

where $B\left(b, b^+\right)$, $B^+\left(b, b^+\right)$ are the arbitrary functions of bosonic operators.

The 2×2 matrix relation is often used in construction of these operators. Since n_F takes only two values $(n_F = 0,1)$, it is convenient to choose the representation for eigenstates in two–component form

$$\psi = \begin{pmatrix} \psi_1 \\ \psi_0 \end{pmatrix}$$

where the upper component ψ_1 corresponds to $n_F = 1$, while ψ_0 - to $n_F = 0$. The fermionic operators are taken on the form of 2×2 matrices

$$f^+ = \sigma^+ = \begin{pmatrix} 0 & 1 \\ 0 & 0 \end{pmatrix}$$

$$f = \sigma^- = \begin{pmatrix} 0 & 0 \\ 1 & 0 \end{pmatrix}$$

Then Hamiltonian $\widetilde{H}$ takes the form in this representation

$$\widetilde{H} = \tfrac{1}{2}\{B, B^+\} + \tfrac{1}{2}[B, B^+]\sigma_3$$

If this Hamiltonian is to be quadratic in momentum p, than B - operators take form:

$$B = \tfrac{1}{\sqrt{2}}(ip + W(x)),\ p = -i\frac{d}{dx}$$

$$B^+ = \tfrac{1}{\sqrt{2}}(-ip + W(x)),$$

Here $W(x)$ is an arbitrary function of coordinate. In this representation SUSY Hamiltonian takes the form

$$\widetilde{H} = \tfrac{1}{2}\left[p^2 + W^2(x) + \sigma_3 W'(x)\right]$$

It is the Witten's Hamiltonian. $W(x)$ is called as "superpotential".

In the 2×2 form

$$\widetilde{H} = \begin{pmatrix} H_+ & 0 \\ 0 & H_- \end{pmatrix}$$

where $H_+ = BB^+$ and $H_- = B^+B$. They are called the partner Hamiltonians. In the explicit form

$$H_\pm = \frac{p^2}{2} + V_\pm(x) \tag{IV.6}$$

where

$$V_\pm(x) = \frac{1}{2}\left[W^2 \mp W'(x)\right] \tag{IV.7}$$

are partner potentials.

It is not difficult to convince that zero energy eigenvalue problem reduces to equations

$$\left[\frac{d}{dx} \mp W(x)\right]\psi_\pm = 0$$

Solutions of which has the form

$$\psi_\pm = c\exp\left[\pm\int_0^x W(x')dx'\right] \tag{IV.8}$$

In order the functions $\psi_\pm$ are really eigenfunctions of $\widetilde{H}$, they must be quadratically integrable. For this it is necessary fulfillment of the following conditions:

For normalizability for

$$\psi_-: \quad \int_0^x W(x')dx' \to +\infty, \qquad when\ \ x \to \pm\infty$$

and that for

$$\psi_+: \quad \int_0^x W(x')dx' \to -\infty, \qquad when\ \ x \to \pm\infty$$

We see that these two conditions contradict each others therefore only one of $\psi_\pm$ could be normalizable. While it may be that no ones of them is normalizable.

So, we conclude that if there is a state with $E = 0$ zero energy, then it is not degenerate and only one function from $\psi_\pm$ corresponds to it.

Recent studies of quantum mechanics show that for any Hamiltonian with one degree of freedom, a partner Hamiltonian can always be constructed such that the resulting system as a whole is supersymmetric.

After such a brief discussion of SUSY quantum mechanics let us consider the real physical example where SUSY is present. It is a problem of electron's motion in the uniform magnetic field $\vec{B} = const$, which is perpendicular of motion's plane. It is known from textbooks that energies of levels (Landau levels) are given by:

$$E = \frac{eB}{m}\left(n + \tfrac{1}{2}\right) \pm \tfrac{1}{2}\frac{eB}{m} = \frac{eB}{m}\left[\left(n_B + \tfrac{1}{2}\right) \pm \left(n_F - \tfrac{1}{2}\right)\right] \qquad \text{(IV.9)}$$

This equality has the form of the supersymmetric harmonic oscillator (Compare with Eq. (IV.4)). "Bosonic" oscillator appears as a result of quantization of electron's orbital motion in a magnetic field, while "fermionic" part describes the interaction of electron's magnetic moment with the field, the frequencies of these two motion coincide to each others, therefore all levels, except one, are two-fold degenerates. It is interesting that SUSY takes place only in case, when the magnetic moment of electron equals to $-e\hbar/2me$, the value, which predicts only the Dirac theory [6].

In this problem the role of fermionic degrees of freedom plays the spin of electron. In case of SUSY transformation electron's spin changes its direction on the reverse one and at the same time it passes from one orbit to another so that its energy remains unchanged.

Below physically more interesting problems will be examined and will be shown that the hidden dynamical symmetries considered in classical mechanics manifest themselves as supersymmetries.

IV.2. Supersymmetry and the Radial Problem

To formulate SUSY in three dimensions for central potentials one generally subjects the radial part of the Schroedinger equation to a similar one-dimensional treatment [1-3]. The radial equation has the form:

$$-\frac{1}{2r^2}\frac{d}{dr}\left(r^2\frac{dR}{dr}\right) + \left[V(r) - E + \frac{l(l+1)}{2r^2}\right]R = 0 \qquad \text{(IV.10)}$$

in which the first order derivative therm can be removed by making a further transformation $R \to u(r)/r$.

As a result, (IV.10) can be reduced to

$$-\frac{1}{2}\frac{d^2u}{dr^2}+\left[V(r)-E+\frac{l(l+1)}{2r^2}\right]u(r)=0$$

this is in the form of a Schroedinger equation similar to that of the one-dimensional problems and can be subjected to a supersymmetric treatment. However, for the radial equation $r\in(0,\infty)$ it constitutes only a half-line problem.

Let consider the Coulomb problem $V(r)=-Ze^2/r$. The radial equation now is

$$\left(-\frac{1}{2}\frac{d^2}{dr^2}-E-\frac{1}{r}+\frac{l(l+1)}{2r^2}\right)u_{nl}(r)=0 \tag{IV.11}$$

Here the scaled variable is used $r\to \frac{1}{mZe^2}r$. According to definitions (IV.6-7) we find

$$\frac{1}{2}\left(W^2-W'\right)=V-E_0=-\frac{1}{r}+\frac{l(l+1)}{2r^2}+\frac{1}{2(l+1)^2} \tag{IV.12}$$

Inspection reveals the relation of this equation to be

$$W(r)=\frac{1}{l+1}-\frac{l+1}{r} \tag{IV.13}$$

Then the supersymmetric partner to V_+ be

$$V_-\equiv\frac{1}{2}\left(W^2+W'\right)=-\frac{1}{r}+\frac{(l+1)(l+2)}{2r^2}+\frac{1}{2(l+1)^2} \tag{IV.14}$$

Now consider the case: <u>*fixed n but a variable l*</u>. Whereas the Bohr series for (IV.12) starts from $(l+1)$ with energies

$$\frac{1}{2}\left[\frac{1}{(l+1)^2}-\frac{1}{n^2}\right]$$

We see from (IV.14) that the lowest level for V_- begins at $n=l+2$ with $n\geq l+2$. Thus the spectrum of H_+ and H_- can be used to give a SUSY interpretation of the hydrogen atom spectrum which corresponds to the well-known hydrogenic ($ns-np$)

degeneracy. To see this point more clearly, we may set $l = 0$. Then V_+ describes the ns states with $n \geq 1$ while V_- corresponds to np levels with $n \geq 2$.

In this way we established a supersymmetric connection between atoms in three dimensions. SUSY brings out a connection between states of same n but different l.

IV. 3. Exact Supersymmetry in the Non–Relativistic Hydrogen Atom

The hydrogen atom (Kepler–Coulomb) problem was considered in the previous Chapters both in classically and quantum mechanically. We saw that the hidden symmetry (LRL vector) had a decisive place in obtaining of orbit equation (classically) or energy levels (quantum mechanically). The accidental degeneracy of spectra takes place in quantum mechanics. Now we'll see that this degeneracy is provided by the conserved LRL vector.

To construct supercharges for the Coulomb problem we must include the spin degrees of freedom. It is an interesting fact that it is easier to discuss the motion in a Coulomb field of a Pauli particle (non–relativistic spin - $1/2$ particle with kinematically independent spin) than the motion of spinless particle. This simplicity results from using the spin operator $\vec{\sigma}$, together with the two vector operators $\vec{L}$ and $\vec{A}$, to construct invariant linear operators that enable one to factorize both angular momentum operator and the radial equation.

Let us sketch briefly how this comes about [7]:

The energy eigenststes are now of the form

$$\left|nlm\right\rangle \otimes \left|\tfrac{1}{2}\mu\right\rangle$$

or equivalently, they may be taken to be

$$R_{ne}(r)\vec{Y}_{jm}^{(e1/2)}$$

where $\vec{Y}_{jm}^{(e1/2)}$ are the Pauli central field spinors.

Let us introduce now the Dirac's quantum number k. This number is the eigenvalue of the scalar operator K defined by [7]

$$K = -(\vec{\sigma}\cdot\vec{L}+1) = -(2\vec{S}\cdot\vec{L}+1) = \vec{L}^2 - \vec{J}^2 - \frac{1}{4} \qquad \text{(IV.15)}$$

Using this definition, one finds that the orbital angular momentum $\vec{L}$ and the total angular momentum $\vec{J} = \vec{L} + 1/2\,\vec{\sigma}$ are each expressible as a quadratic form in K:

$$\vec{L}^2 = k(k+1), \quad \vec{J}^2 = K^2 - 1/4 \tag{IV.16}$$

One next finds that the eigenvalue k of K may be any integer, *zero excluded*:

$$k = ..., -2, -1, 1, 2, ...$$

For a prescribed value of k the quantum numbers l and j are then obtained as

$$\begin{aligned} l &= l(k) = |k| + \tfrac{1}{2}(-1 + \operatorname{sgn} k), \\ j &= j(k) = |k| - \tfrac{1}{2} \end{aligned} \tag{IV.17}$$

where $\operatorname{sgn} k$ denotes the sign of k.

A Z_2 - grading can now be introduced in the Hilbert space of states by classifying the states to be even or odd with respect to the parity operator [7]

$$P_k = \frac{K}{|k|} \tag{IV.18}$$

i.e. having eigenvalues $\pm \operatorname{sgn} k$ respectively. Equally, linear operators can be assigned a grade. An operator is even, if it commutes with P_k, whereas operators anticommuting with P_k are called odd.

As an example of an even operator we mention H, following from the fact that K is built from symmetry operators of H. To find odd operators we make use of the following theorem:

Theorem: Suppose $\vec{V}$ is a vector with respect to the orbital angular momentum $\vec{L}$ that is also perpendicular to $\vec{L}$

$$\begin{aligned} &[L_i, V_j] = i\varepsilon_{ijk} V_k \\ &\vec{L} \cdot \vec{V} = \vec{V} \cdot \vec{L} = 0 \end{aligned}$$

Then K anticommutes with the $\vec{J}$-scalar $\vec{S} \cdot \vec{V}$.

This theorem supplies us with odd operators $\vec{S} \cdot \vec{V}$ with $\vec{V}$ equal to, for example $\vec{p}, \vec{r}$ *or* $\vec{A}_0$.

Considering the square of the odd operator $\vec{S} \cdot \vec{A}_0$. It is readly shown that [7]

$$\left(\vec{S}\cdot\vec{A}_0\right)^2=\frac{1}{2}HK^2+\frac{1}{4}$$

We now define the supercharge

$$Q_1\equiv\frac{\vec{S}\cdot\vec{A}_0}{k} \qquad \text{(IV.19)}$$

and the shifted Hamiltonian (Witten's Hamiltonian)

$$\widetilde{H}=H+\frac{1}{2k^2} \qquad \text{(IV.20)}$$

Square of Q_1^2 can then be identified with the $S(2)$ product–rule

$$\{Q_1,Q_1\}=\widetilde{H}$$

Moreover, using the second charge as

$$Q_2\equiv iQ_1P_k=\frac{i}{k^2}\left(\vec{S}\cdot\vec{A}_0\right)K\,, \qquad \text{(IV.21)}$$

thereby completing the $S(2)$ symmetry algebra.

Instead of working in the Hermitian representation of $S(2)$ as described above, one often uses the odd ladder operators

$$Q_\pm=\frac{1}{\sqrt{2}}\left(Q_1\pm iQ_2\right)=\frac{1}{\sqrt{2}k}\left(\vec{S}\cdot\vec{A}_0\right)\left(1\mp P_k\right) \qquad \text{(IV.22)}$$

Note that for the even states $Q_+=0$ and $Q_-=\dfrac{\sqrt{2}\vec{S}\cdot\vec{A}_0}{k}$, whereas for the odd states their roles are reversed.

We now turn to discuss the implications of the constructed $S(2)$ symmetry on the spectrum. The consequence of a symmetry group of a Hamiltonian is that the energy eigenspaces consist of irreducible representations of it. The irreducible representations of $S(2)$ are known to be either one– or two–dimensional. Consider the energy eigenstates of the Pauli particle, given by

$$\phi_{El(k)j(k)m}\left(r,\theta,\varphi\right)=R_{El(k)}\left(r\right)Y^{\left(l(k)\frac{1}{2}\right)j(k)m}\left(\theta,\varphi\right)$$

The quantum number k assumes the values $\pm k$. Applying the ladder operators (IV.22) to this equation yields either zero or $\frac{\sqrt{2}\vec{S}\cdot\vec{A}_0}{k}$. Moreover, since $\{K,\vec{S}\hat{\vec{r}}\}=0$ (by theorem) and $\left(\vec{S}\cdot\hat{\vec{r}}\right)^2=\frac{1}{4}$, we obtain the nice property that

$$\left(\vec{S}\cdot\hat{\vec{r}}\right)\chi_m^{\pm k}=-\frac{1}{2}\chi_m^{\mp k}$$

Therefore we find

$$\left(\frac{\sqrt{2}\vec{S}\cdot\vec{A}_0}{k}\right)\phi_{E\pm km}=-\left(E+\frac{1}{2k^2}\right)^{1/2}\phi_{E\mp km} \quad \text{(IV.23)}$$

Hence states within the subspace H_k, with fixed E and m, transform irreducibly under the superalgebra $S(2)$. The multiplets have dimensionality two, unless $E=-\frac{1}{2k^2}$, in which case they are one–dimensional.

The previous equation (IV.23) can be reduced to a set of radial equations. Using the identity

$$\vec{S}\cdot\vec{A}_0=\left[1-k\left(i\hat{\vec{r}}\cdot\vec{p}+\frac{K+1}{r}\right)\right]\left(\vec{S}\cdot\hat{\vec{r}}\right)$$

and substituting in the previous equation (IV.23), we find $\left(k=\pm|k|\right)$

$$\frac{1}{\sqrt{2}k}\left[1-k\left(\frac{d}{dr}+\frac{k+1}{r}\right)\right]R_{El(k)}(r)=\left(E+\frac{1}{2k^2}\right)^{\frac{1}{2}}R_{El(-k)}(r) \quad \text{(IV.24)}$$

Introducing furthermore $rR=\chi$ and identifying $k=l+1$, these two equations can be written as

$$\begin{pmatrix}0 & A^-(l)\\ A^+(l) & 0\end{pmatrix}\begin{pmatrix}\chi_{El}(r)\\ \chi_{El+1}(r)\end{pmatrix}=\sqrt{E+\frac{1}{2(l+1)^2}}\begin{pmatrix}\chi_{El}(r)\\ \chi_{El+1}(r)\end{pmatrix} \quad \text{(IV.25)}$$

where

$$A^{\pm}(l) = \frac{1}{\sqrt{2}}\left[\pm\frac{d}{dr} - \frac{l+1}{r} + \frac{1}{l+1}\right] \tag{IV.26}$$

The radial ladder operators $A^{\pm}(l)$ are identical to those found in the radial SUSY studies of the Coulomb problem ([3], and the previous Section). We stress the fact that in those cases the angular momentum is not really affected, because the supercharges are purely radial. In the present three–dimensional study the l-value is changed by the charges $(\pm k \to \mp k)$.

We have used the operators $\vec{A}_0, \vec{L}$ and $\vec{S}$ to construct the supercharges Q_1 and Q_2 of a $S(2)$ supersymmetry algebra of the non-relativistic Coulomb problem. This supersymmetry occurs for all levels with fixed j, m, as opposed to the $SO(4)$ which only concerns the bound states. The results derived show that SUSY in Coulomb problem can be extended beyond the one–dimensional formulation. The symmetry and spectrum–generating algebra of such a system are identical to those of the hydrogen atom. It therefore seems reasonable to expect the $S(2)$ supersymmetry to hold as well.

It is very remarkable also that the supercharge generator Q_1 in this problem is related to the LRL vector, it is simply the projection of LRL vector $\vec{A}_0$ on the direction of electron spin $\vec{S}$. But in this non-relativistic case spin is a kinematically independent quantity.

REFERENCES

[1] Weinberg S. The Quantum Theory of Fields, Vol.III, Supersymmetry. Cambridge Univ. Press, 2000.

[2] Gendenshtein L.E. Krive I.V. Supersymmetry in Quantum Mechanics, *Sov.Phys. Usp.*, 28, 645 (1985)

[3] Lahiri A., Roy P.K., Bagchi B., Supersymmetry in Quantum Mechanics. *Int. J. Mod. Phys.* A5, 1383 (1990).

[4] Bagchi B.K., Supersymmetry in Quantum and Classical Mechanics. ChapmanandHall/CRC, Boca Baton etc. 2000.

[5] Witten E. Spontaneous Symmetry Breaking in SUSY Quantum Mechanics. *Nucl.Phys.* B188, 513 (1981).

[6] Dirac P.A.M., Principles of Quantum Mechanics, Cambridge Univ. Press, Cambridge, 1932.

[7] Biedenharn L.C., Louck L.D. Angular Momentum in Quantum Physics. Encyclopedia of Mathematics and its Application", Addison-Wisley Pub. Company, 1981.

[8] Tangerman R.D., Tjon J.A. Exact Supersymmetry in the Nonrelativistic Hydrogen Atom, *Phys.Rev.,* A48, 1089 (1993).

Chapter V

RELATIVISTIC QUANTUM MECHANICS

V.1. SUPERSYMMETRY IN THE DIRAC EQUATION FOR THE COULOMB POTENTIAL

In 1916 Arnold J.W. Sommerfeld [1] applied the quantization rules of the old quantum theory to the relativistic hydrogen atom. He obtained an elegant closed form as his answer for the bound state energy levels . Ten years later when the new Schroedinger wave mechanics replaced the *ad hoc* rules of the Bohr quantum theory, Sommerfeld's result was regarded as something of an anachronism , despite the fact that his formula agreed very well indeed with experimental data.

When Dirac [2] developed relativistic quantum mechanics, the relativistic Coulomb problem proved to be *exactly solvable.* But the resulting formula for the energy levels was truly a surprise: *the new answer* was precisely the old Sommerfeld formula. The secret of this puzzle was revealed by L.C.Biedenharn [3]. He found, that the underlying symmetry of the problem intervenes in a most remarkable and essential way so as to produce the closest possible correspondence between the two calculations. The details of this concept will be discussed in one of sequential Chapter. Now let us return to the direct problem.

The Sommerfeld formula for the hydrogen atom spectrum reads

$$E = m\left\{1+\frac{(Z\alpha)^2}{\left(n-|\kappa|+\sqrt{\kappa^2-(Z\alpha)^2}\right)}\right\}^{-1/2}$$

Here κ is the eigenvalue of the Dirac operator

$$K = \beta\left(\vec{\Sigma}\cdot\vec{l}+1\right)$$

which commutes with the Dirac Hamiltonian

$$H = \vec{\alpha}\cdot\vec{p}+\beta m-\frac{a}{r}, \qquad a \equiv Ze^2 = Z\alpha$$

$\vec{l}$ is the angular momentum vector, $\vec{\alpha}$ and β are usual Dirac matrices and $e^2 = \alpha$ is the fine structure constant, while $\vec{\Sigma}$ is the electron spin matrix $\vec{\Sigma} = diag(\vec{\sigma}, \vec{\sigma})$.

Eigenvalue $|\kappa| = j + \frac{1}{2}$. Let us mention the degeneracy of spectrum with respect to signs of κ.

It is surprising that for other solvable potentials the degeneracy with respect to signs of κ does not take place.

Physically this degeneracy leads to the forbidden of the Lamb shift. Indeed, positive $\kappa = j + \frac{1}{2}$ corresponds to alined spin $j = l + \frac{1}{2}$, i.e. to states $(S_{\frac{1}{2}}, P_{\frac{3}{2}}, etc.)$, while negative $\kappa = -(j + \frac{1}{2})$ corresponds to unalined spin $j = l - \frac{1}{2}$, i.e. to states $(p_{\frac{1}{2}}, d_{\frac{3}{2}}, etc.)$. So the absence of the Lamb shift $\left(E_{S_{\frac{1}{2}}} - E_{P_{\frac{1}{2}}}\right)$ is a consequence of above mentioned degeneracy $\kappa \to -\kappa$.

As we saw in the case of non–relativistic Pauli electron the cooperative effect of hidden symmetry and the doubling of levels owing to spin directions give us a new $S(2)$ symmetry $(N = 2\ \ SUSY)$ for the hydrogen atom. Therefore it's natural to ask whether is there a supersymmetry in the Dirac equation for the Coulomb potential, where we have both the same potential and spin, which is present on the natural ground.

Now one can convinced that the Dirac equation for the Coulomb problem can be solved algebraically [2]. Separating the angular variables in the usual manners, one can write the Coulomb radial equations as follows:

$$G_k'(r) + \frac{k}{r} G_k - (\alpha_1 - V) F_k = 0$$

$$F_k'(r) - \frac{k}{r} F_k - (\alpha_2 + V) G_k = 0$$

where $\alpha_1 = m + E$, $\alpha_2 = m - E$ and G_k is the "up" component, which is "large" in non–relativistic limit, while F_k is the down component. Now substitute the Coulomb potential

$$V = -\frac{a}{r}$$

and rewrite the above coupled equations in the matrix form

$$\begin{pmatrix} G_k'(r) \\ F_k'(r) \end{pmatrix} + \frac{1}{r} \begin{pmatrix} k, & -a \\ a, & -k \end{pmatrix} \begin{pmatrix} G_k \\ F_k \end{pmatrix} = \begin{pmatrix} 0 & \alpha_1 \\ \alpha_2 & 0 \end{pmatrix} \begin{pmatrix} G_k \\ F_k \end{pmatrix} \tag{V.1}$$

Following Sukumar [2] we notice that the matrix multiplying $1/r$ can be diagonalized by a matrix D, where

$$D = \begin{pmatrix} k+s, & -a \\ -a, & k+s \end{pmatrix} \tag{V.2}$$

$$D^{-1} = \frac{a}{(k+s)^2 - a^2} \begin{pmatrix} \frac{k+s}{a}, & 1 \\ 1, & \frac{k+s}{a} \end{pmatrix}, \quad s = \sqrt{k^2 - a^2}$$

On multiplying above equation from the left by the matrix D and introducing the new variable $\rho = Er$ leads to the pair of equations

$$A\widetilde{F} = \left(\frac{m}{E} - \frac{k}{s}\right)\widetilde{G},$$

$$A^{+}\widetilde{G} = -\left(\frac{m}{E} + \frac{k}{s}\right)\widetilde{F},$$

where

$$\begin{pmatrix} \widetilde{G} \\ \widetilde{F} \end{pmatrix} = D \begin{pmatrix} G \\ F \end{pmatrix}$$

and

$$A = \frac{d}{d\rho} - \frac{s}{\rho} + \frac{a}{s},$$

$$A^{+} = -\frac{d}{d\rho} - \frac{s}{\rho} + \frac{a}{s}$$

Thus we can easily decouple the equations for $\widetilde{F}$ and $\widetilde{G}$ thereby obtaining

$$H_{-}\widetilde{F} \equiv A^{+}AF = \left(\frac{k^2}{s^2} - \frac{m^2}{E^2}\right)\widetilde{F}$$

$$H_{+}\widetilde{G} \equiv AA^{+}\widetilde{G} = \left(\frac{k^2}{s^2} - \frac{m^2}{E^2}\right)\widetilde{G}$$

We thus see that there is a supersymmetry in the problem and $H_{\pm}$ are shape invariant supersymmetric partner potentials since

$$H_{+}(\rho;s,a)=H_{-}(\rho;s+1,a)+\frac{a^{2}}{s^{2}}-\frac{a^{2}}{(s+1)^{2}} \tag{V.3}$$

On comparing with the traditional shape invariant notations (see, Appendix, below) we have in our case

$$a_{2}=s+1,\;\; a_{1}=s,\quad R(a_{2})=\frac{a^{2}}{a_{1}^{2}}-\frac{a^{2}}{a_{2}^{2}}$$

So that the energy eigenvalues of H are given by

$$\frac{k^{2}}{s^{2}}-\frac{m^{2}}{E_{n}^{2}}\equiv E_{n}^{(-)}=\sum_{k=1}^{n+1}R(a_{k})=a^{2}\left(\frac{1}{s^{2}}-\frac{1}{(s+1)^{2}}\right)$$

Thus the Coulomb bound state energy eigenvalues E_{n} are

$$E_{n}=\frac{m}{\left[1+\frac{a^{2}}{(s+n)^{2}}\right]^{1/2}},\qquad n=0,1,2,... \tag{V.4}$$

It should be noted that every eigenvalue of H_{-} is also an eigenvalue of H_{+} except for the ground state of H_{-} which satisfy

$$A\tilde{F}\Rightarrow\tilde{F}_{0}(\rho)=\rho^{s}\exp\left(-{}^{a\rho}\!/\!_{s}\right)$$

Using the formalism for shape invariant Hamiltonian, one can also algebraically obtain all the eigenfunctions of $\tilde{F}$ and $\tilde{G}$.

Notice that spectrum only depends on $|k|$ leading to a doublet of states corresponding to $k=|k|$ and $k=-|k|$ for all positive n. However, for $n=0$, only the negative value of k is allowed and hence this is a singlet state.

APPENDIX: SHAPE INVARIANCE (SI)

Let us now explain precisely what one means by shape invariance. If the pair of SUSY partner potentials $V_{\pm}(x)$ defined in previous Section are similar in shape and differ only in the parameters that appear in them, then they are said to be shape invariant. More precisely, if the profiles of $V_{\pm}(x)$ are such that they satisfy the relationship

$$V_{-}(x,c_0)=V_{+}(x,c_1)+R(c_1) \tag{A.1}$$

where the parameter c_1 is some function of c_0, say given by $c_1=f(c_0)$, the potentials $V_{\pm}$ are said to be SI. In other words, to be SI the potentials $V_{\pm}$ while sharing a similar coordinate dependence can at most differ in the presence of some parameters.

An example will make the definition of SI clear. Let us take

$$W(x)=c_0\tanh x: \quad W(\infty)=-W(-\infty)=c_0 \tag{A.2}$$

Then

$$V_{\pm}(x,c_0)=-\frac{1}{2}c_0(c_0\pm 1)\operatorname{sec}h^2 x+\frac{c_0^2}{2} \tag{A.3}$$

But these can also be expressed as

$$V_{-}(x,c_0)=V_{+}(x,c_0)+R(c_1); \quad c_1=c_0-1 \tag{A.4}$$

where

$$R(c_1)=\frac{1}{2}\left(c_0^2-c_1^2\right) \tag{A.5}$$

So the potentials $V_{\pm}$ are SI in accordance with definition.

To exploit the SI condition let us assume that (A.1) holds for a sequence of parameters $\{c_k\}$, $k=0,1,2,\ldots$ where $c_k=f^k(c_0)$

Then

$$H_{-}(x,c_k)=H_{+}(x,c_{k+1})+R(c_k) \tag{A.6}$$

and we call $H^{(0)}=H_{+}(x,c_0)$, $H^{(1)}=H_{-}(x,c_0)$. Writing $H^{(m)}$ as

$$H^{(m)} = -\frac{1}{2}\frac{d^2}{dx^2} + V_+(x, c_m) + \sum_{k=1}^{m} R(c_k) = H_+(x, c_m) + \sum_{k=1}^{m} R(c_k) \tag{A.7}$$

It follows on using (A.2) that

$$H^{(m+1)} = H_-(x, c_m) + \sum_{k=1}^{m} R(c_k) \tag{A.8}$$

Thus we are able to set up a hierarchy of Hamiltonians $H^{(m)}$ for various m values. Now according to the principles of SUSY quantum mechanics, say H_+ contains the lowest state with a zero–energy eigenvalue. It then transpires from (A.7) that the lowest energy level of $H^{(m)}$ has the value

$$E_0^{(m)} = \sum_{k=1}^{m} R(c_k) \tag{A.9}$$

It is not too difficult to realize that because of the chain

$$H^{(m)} \to H^{(m-1)} \to \ldots \to H^{(1)}(\equiv H_-) \to H_0(\equiv H_+),$$

the nth member in this sequence carries the nth level of the energy spectra of $H^{(0)}$ (or H_+), namely

$$E_n^{(+)} = \sum_{k=1}^{n} R(c_k), \quad E_0^{(+)} = 0 \tag{A.10}$$

Let us now return to the example above. We rewrite

$$V_-(x, c_0) = V_+(x, c_0 - 1) + \frac{1}{2}c_0^2 - \frac{1}{2}(c_0 - 1)^2 \tag{A.11}$$

and note that we can generate c_k from c_0 as $c_k = c_0 - k$. Therefore the levels of $V_+(x, c_0)$ are given by

$$E_n^+ = \sum_{k=1}^{n} R(c_k) = \frac{1}{2}\sum_{k=1}^{n}\left(c_0^2 - c_k^2\right) = \frac{1}{2}\left(c_0^2 - c_n^2\right) = \frac{1}{2}\left[c_0^2 - (c_0 - n)^2\right] \tag{A.12}$$

(A.2) corresponds to the potential (see, (A.3))

$$V_+(x) = -\beta \sec h^2 x = -\frac{\beta}{\cosh^2 x} \tag{A.13}$$

where $\beta = \frac{1}{2} c_0 (c_0 + 1)$. We find

$$E_n = E_n^+ - \frac{1}{2} c_0^2 = -\frac{1}{2} (c_0 - n)^2 \tag{A.14}$$

Let us remark that we have used a supersymmetry to solve problem, which in its original formulation did not contain this symmetry. The extension of this problem, i.e. the passing to SUSY quantum mechanics seems to be very useful.

V. 2. An "Accidental Symmetry" Operator for the Dirac Equation in the Coulomb Potential – from Pauli to Dirac

We saw that well known l degeneracy of hydrogen atom spectrum in non – relativistic quantum mechanics is removed in case of the Dirac equation (since LRL vector is no longer conserved in this case). However the degeneracy is not lifted completely even in this case. The two–fold degeneracy in total angular momentum still remains and it contradicts with existence of well known experimental result–the Lamb shift.

The analogue of LRL vector in the Dirac equation is the Johnson – Lippmann (JL) operator which in 1950 was published by these authors in the form of brief abstract [3]. Because of importance and rarity of this problem, below we display the full text of the above mentioned abstract as it was presented then.

"J6. *Relativistic Kepler Problem.* M.H. Johnson and B.A. Lippmann, *Naval Resarch Laboratory* – Besides the usual integrals of motion, $\vec{M}$ and j (in Dirac's notation) the relativistic equations for a charge in a Coulomb field admit

$$A = \vec{\sigma} \cdot \vec{r} \cdot r^{-1} - i\left(\frac{\hbar c}{e^2}\right)\left(mc^2\right)^{-1} j\rho_1\left(H - mc^2 \rho_3\right)$$

as another integral of motion. Since A and j anticommute, the pairs with same $|j|$ are degenerate. Thus the existence of A establishes the "accidental" degeneracy in relativistic Kepler problem just as the existence of the axial vector establishes the degeneracy with respect to l in the corresponding non–relativistic problem."

As regards of derivation of this operator to our surprise it was not published anywhere in scientific literature (one of the curious fact in the history of physics of 20th century), nevertheless there were many publications, where this operator was applied.

The JL operator were generalized even to arbitrary dimensions [4] and it was shown that this operator can be used to construct relativistic supercharge in Witten's superalgebra. As far as commutativity of JL operator with the Dirac Hamiltonian is concerned, it was only mentioned that it can be proved by "rather tedious manipulations" [4].

Below we construct this operator in rather simple and very transparent and understandable way [5].

First of all, we return to the introductory relations from Section V.1.

It is well known that the Dirac operator $K = \beta(\vec{\Sigma}\vec{l} + 1)$ commutes with the Dirac–Coulomb Hamiltonian

$$H = \vec{\alpha} \cdot \vec{p} + \beta m - \frac{a}{r}, \quad a \equiv Ze^2 = Z\alpha$$

(In JL abstract K is denoted as j). We follow the simple logic:

- If there is some symmetry that relates two signs of κ, the corresponding symmetry operator must anticommute with K.
- At the same time, naturally, this operator has to commute with the Dirac Hamiltonian.

Thus the first step will be constructing an operator(s), that anticommute with K. In order to find such an operator, let generalize the theorem, which was known earlier for the Pauli electron (See Sec IV.2). We reformulate it in the following form [5, 6]:

Theorem: Let $\vec{V}$ be a vector with respect to the angular momentum $\vec{l}$, i.e.

$$[l_i, V_j] = i\varepsilon_{ijk} V_k$$

In the vector product form it can be written as

$$\vec{l} \times \vec{V} + \vec{V} \times \vec{l} = 2i\vec{V}.$$

Suppose also that this vector is perpendicular to $\vec{l}$

$$(\vec{l} \cdot \vec{V}) = (\vec{V} \cdot \vec{l}) = 0$$

Then K anticommutes with operator $(\vec{\Sigma} \cdot \vec{V})$, which is scalar with respect of the total $\vec{J} = \vec{l} + \frac{1}{2}\vec{\Sigma}$ momentum.

Proof:

Let us consider a product $(\vec{\Sigma}\cdot\vec{l})(\vec{\Sigma}\cdot\vec{V})$. Exploiting the known properties of Dirac matrices and conditions of the theorem, one can establish that

$$(\vec{\Sigma}\cdot\vec{l})(\vec{\Sigma}\cdot\vec{V}) = \left(\vec{l}\cdot\vec{V}\right)+i\left(\vec{\Sigma},\, \vec{l}\times\vec{V}\right)= i\left(\vec{\Sigma},\, 2i\vec{V}-\vec{V}\times\vec{l}\right)= \\ = -2\left(\vec{\Sigma}\cdot\vec{V}\right)-i\left(\vec{\Sigma},\, \vec{V}\times\vec{l}\right)$$

Therefore

$$(\vec{\Sigma};\, \vec{l}+1)(\vec{\Sigma}\cdot\vec{V}) = -\left\{\left(\vec{\Sigma}\cdot\vec{V}\right)+i\left(\vec{\Sigma},\, \vec{V}\times\vec{l}\right)\right\} \tag{V.5}$$

Now consider the same product in reversed order

$$(\vec{\Sigma}\cdot\vec{V})\left(\vec{\Sigma}\cdot\vec{l}\right)=\left(\vec{V}\cdot\vec{l}\right)+i\left(\vec{\Sigma},\, \vec{V}\times\vec{l}\right)=i\left(\vec{\Sigma},\, \vec{V}\times\vec{l}\right)$$

Hence

$$\left(\vec{\Sigma}\cdot\vec{V}\right)\left(\vec{\Sigma}\cdot\vec{l}+1\right)=\left(\vec{\Sigma}\cdot\vec{V}\right)+i\left(\vec{\Sigma},\, \vec{V}\times\vec{l}\right)= \\ = -\left(\vec{\Sigma}\cdot\vec{l}+1\right)\left(\vec{\Sigma}\cdot\vec{V}\right)$$

In the last step we made use of the equation (V.5). Therefore we have obtained

$$\left\{\vec{\Sigma}\cdot\vec{l}+1,\ \vec{\Sigma}\cdot\vec{V}\right\}=0$$

Now, according to the definition of K-operator, it follows that

$$K\left(\vec{\Sigma}\cdot\vec{V}\right)=-\left(\vec{\Sigma}\cdot\vec{V}\right)K \tag{V.6}$$

Thus the theorem is proved.

It is evident that the class of anticommuting with K (so called $K-odd$) operators is not restricted by these operators only – any operator of kind $\hat{O}\left(\vec{\Sigma}\cdot\vec{V}\right)$, where $\hat{O}$ is commuting with K, but otherwise arbitrary, also is $K-odd$.

It is noteworthy, that the following useful relation holds in the framework of constraints of above theorem:

$$K\left(\vec{\Sigma}\cdot\vec{V}\right)=-i\beta\left(\vec{\Sigma},\, \frac{1}{2}\left[\vec{V}\times\vec{l}-\vec{l}\times\vec{V}\right]\right), \tag{V.7}$$

which follows from vector's transformation rule mentioned in theorem in the form of vectorial product.

One can see that the antisymmetrized vector product characteristic of LRL vector appears on the right hand–side of this relation. Important special cases, resulting from the above theorem include $\vec{V} = \hat{\vec{r}}$ (unit radial vector), $\vec{V} = \vec{p}$ (linear momentum) and $\vec{V} = \vec{A}$ (LRL vector). This last one as we know has the following form

$$\vec{A} = \hat{\vec{r}} - \frac{i}{2ma}\left[\vec{p} \times \vec{l} - \vec{l} \times \vec{p}\right] \qquad \text{(V.7)}$$

According to (V.7) there appears a relation between above three odd operators

$$\vec{\Sigma} \cdot \vec{A} = \vec{\Sigma} \cdot \hat{\vec{r}} + \frac{i}{ma}\beta K\left(\vec{\Sigma} \cdot \vec{p}\right) \qquad \text{(V.8)}$$

At this point we are tooled up to derive the hidden symmetry operator (the second step).

The second step of our two-stage derivation strategy is to find $K - odd$ operators that commute with the Dirac–Coulomb Hamiltonian. There remains still considerable freedom of choice here because of above mentioned remark about operators like $\hat{O}\left(\vec{\Sigma} \cdot \vec{V}\right)$. One can take $\hat{O}$ into consideration or ignore it.

Let us choose

$$\vec{\Sigma} \cdot \hat{\vec{r}} \text{ and } K\left(\vec{\Sigma} \cdot \vec{p}\right) \qquad \text{(V.9)}$$

Because of specific place of LRL vector, it is reasonable to assume that $\left(\vec{\Sigma} \cdot \vec{A}\right)$ - like term will present here. So this choice is dictated by (V.8). Both operators in (V.9) are diagonal matrices. After commuting them with non – diagonal H we will eventually end up with non–diagonal terms. For example,

$$\left[\vec{\Sigma} \cdot \hat{\vec{r}}, H\right] = \frac{2i}{r}\beta K \gamma^5$$

(See, Appendix below). The resulting matrix on the right-hand-side is anti– diagonal. Let probe the following combination of diagonal and anti-diagonal matrices

$$A' = x_1\left(\vec{\Sigma} \cdot \hat{\vec{r}}\right) + ix_2 K\left(\vec{\Sigma} \cdot \vec{p}\right) + ix_3 K\gamma^5 f(r)$$

Here the coefficients are chosen in such a way, that A' be Hermitian, with $x_{1,2,3}$ are arbitrary real numbers and $f(r)$ is an arbitrary scalar function to be determined by the symmetry requirement.

The commutator of A' and H is given by (See, Appendix for details)

$$[A',H] = x_1 \frac{2i}{r}\beta K\gamma^5 - x_2 \frac{a}{r^2} K(\vec{\Sigma}\cdot\hat{\vec{r}}) - \\ - x_3 f'(r) K(\vec{\Sigma}\cdot\hat{\vec{r}}) - i x_3 2m\beta K\gamma^5 f(r)$$

Grouping diagonal and anti–diagonal matrices separately and equating this expression to zero, we obtain equation

$$K(\vec{\Sigma}\cdot\hat{\vec{r}})\left(\frac{a}{r^2}x_2 + x_3 f'(r)\right) + 2i\beta K\gamma^5\left(\frac{1}{r}x_1 - mf(r)x_3\right) = 0$$

This equation is satisfied, if diagonal and anti-diagonal terms become zero separately, i.e.

$$\frac{x_1}{r} = mf(r)x_3 \qquad \frac{a}{r^2}x_2 = -f'(r)x_3$$

Integration of the second equation over the interval (r,∞) yields

$$x_3 f(r) = -\frac{a}{r}x_2$$

and using this result in the first equation, we obtain that

$$x_2 = -\frac{1}{ma}x_1$$

$$x_3 f(r) = \frac{1}{mr}x_1$$

Taking into account these solutions into A' we derive the following operator that commutes with the Dirac Hamiltonian

$$A' = x_1\left\{(\vec{\Sigma}\cdot\hat{\vec{r}}) - \frac{i}{ma}K(\vec{\Sigma}\cdot\vec{p}) + \frac{i}{mr}K\gamma^5\right\}$$

This operator is a $K - odd$ and satisfies all conditions of the above theorem.

One can make sure that A' is a different form of JL operator. Indeed, if we turn to usual $\vec{\alpha}$ matrices making use their definition and taking into account the expression of the Dirac–Coulomb Hamiltonian, A' can be reduced to the form (x_1, as unessential common factor, may be dropped)

$$A' = \gamma^5\left\{\vec{\alpha}\cdot\hat{\vec{r}} - \frac{i}{ma}K\gamma^5(H-\beta m)\right\}$$

This expression is nothing more but the Johnson – Lippmann operator. So we developed the approach which enables us not only quick derivation of Johnson – Lippmann operator but simultaneously its commutativity with the Dirac-Coulomb Hamiltonian is proved.

V. 3. Physical Meaning and Some Applications of Johnson - Lippmann Operator

In order to determine a physical meaning of A' operator note, that using (V.7) this operator may be rewritten in the following form:

$$A' = \vec{\Sigma}\left\{\hat{\vec{r}} - \frac{i}{2ma}(\vec{p}\times\vec{l} - \vec{l}\times\vec{p})\right\} + \frac{i}{mr}K\gamma^5$$

In the non–relativistic limit, when $\beta \to 1$, and $\gamma^5 \to 0$, this operator reduces to the projection of LRL vector on the electron spin direction, $A' \to \vec{\Sigma}\cdot\vec{A}$, or because of $(\vec{l}\cdot\vec{A}) = 0$, it is a projection on the total momentum, $\vec{J}$.

After this it is clear that the Witten algebra can be derived by identification supercharges as

$$Q_1 = A, \quad Q_2 = i\frac{AK}{k} \tag{V.10}$$

It follows that

$$\{Q_1, Q_2\} = 0, \text{ and } Q_1^2 = Q_2^2 = A^2 \tag{V.11}$$

This last factor may be identified as a Witten Hamiltonian ($N = 2$ supersymmetry).

As for further application, let us calculate the square of JL operator. The result is as follows

$$A^2 = 1 + \left(\frac{K}{a}\right)^2\left(\frac{H^2}{m^2} - 1\right) \tag{V.12}$$

Because all operators in this relation commute with each others, one can replace them by their eigenvalues. Therefore one obtains energy spectrum pure algebraically after specifying spectrum of A^2. Since A^2 is positively defined the minimal eigenvalue of A^2 is zero. For this eigenvalue upper relation gives precisely the ground state energy of hydrogen atom,

$$E_0 = m\left(1 - \frac{(Z\alpha)^2}{\kappa^2}\right)^{\frac{1}{2}} \tag{V.13}$$

Full spectrum can be easily derived by well–known ladder procedure described above in the problem of *shape invariance*. This cause to change

$$\sqrt{k^2 - a^2} \to \sqrt{k^2 - a^2} + n - |k| \tag{V.14}$$

It leads to the usual Sommerfeld formula displayed above.

It is worthwhile to note a full analogy with classical mechanics, where closed orbits were derived by calculating the square of LRL vector without solving the differential equation of motion. Therefore we are convinced that the degeneracy of hydrogen atom spectrum with respect to interchange $\kappa \to -\kappa$ is related to the existence of JL operator, which in its turn takes its physical origin from LRL vector.

It is also remarkable that the same symmetry is responsible for absence of the Lamb shift in this problem. Inclusion of the Lamb shift terms,

$$\Delta V_{Lamb} \approx \frac{4\alpha^2}{3m^2}(\ln\frac{m}{\mu} - \frac{1}{5})\delta^3(\vec{r}) + \frac{\alpha^2}{2\pi m^2 r^3}(\vec{\Sigma}\cdot\vec{l})$$

found by calculating radiative corrections to the photon propagator and photon – electron vertex function, into the Dirac Hamiltonian breaks commutativity of A with H. However it is evident that without radiative corrections, terms like ΔV_{Lamb} do not appear in the Dirac Hamiltonian, as in the one–electron theory, and as long as only Coulomb potential is considered, the appearance of the Lamb shift should be always forbidden.

Let us underline that the hidden (dynamical) symmetry, associated to the Coulomb potential, governs a wide range of phenomena from planetary motion till fine and hyperfine structure of atomic spectra.

We can make our attention to one analogy to non-relativistic quantum mechanics. The symmetries $SO(4)$ or $SO(3,1)$ of the bound or scattering states of the non-relativistic Coulomb problem, generated by the angular momentum operator and the appropriately normalized operator corresponding to the LRL vector, do not hold in the Dirac case. The symmetry group generated the constants of motion is no longer uniquely defined. One can introduce the following conserved operators

$$M_1 = \frac{1}{2}\frac{K}{\sqrt{K^2}}, \qquad M_2 \frac{1}{2}\frac{A}{\sqrt{A^2}}, \qquad M_3 = -2iM_1M_2 \tag{V.15}$$

They satisfy the characteristic $SU(2)$ commutation relations

$$\left[M_i, M_j\right] = i\varepsilon_{ijk} M_k$$

and generate together with the total angular momentum $\vec{J}$ the symmetry group $SU(2)\times SU(2)$. This vector $\vec{M} = \left(M_1, M_2, M_3\right)$ has been called "Lenz-spin" [7, 8].

Clearly, the Lenz-spin is conserved spin-1/2 vector operator so that the total symmetry of the Dirac hydrogen atom is $SO(4)$, albeit in a rather trivial way as the solutions always transform as the spin-1/2 representation under the second $SU(2)$ group. Therefore, the solutions of the Dirac hydrogen atom may be classified by the eigenvalues of the complete set of operators: $H,\ \vec{J}^2,\ J_3,\ M_3$.

APPENDIX: CALCULATION OF RELEVANT COMMUTATORS

1) calculation of the commutator $\left[\vec{\Sigma}\cdot\hat{\vec{r}},\, H\right]$:

$$\left[\vec{\Sigma}\cdot\hat{\vec{r}}, \vec{\alpha}\cdot\vec{p} + \beta m + V(r)\right] = \left[\vec{\Sigma}\cdot\hat{\vec{r}},\, \vec{\alpha}\cdot\vec{p}\right] = \Sigma_i\left[\frac{r_i}{r},\, \alpha_j p_j\right] + \left[\Sigma_i, \alpha_j p_j\right]\frac{r_i}{r_j}$$

$$* \ \frac{\Sigma_i \alpha_j}{r}\left[r_i, p_j\right] + \Sigma_i \alpha_j\left[\frac{1}{r}, p_j\right] r_i = \frac{i}{r}\vec{\Sigma}\vec{\alpha} - i\vec{\Sigma}\vec{r}\,\frac{\vec{\alpha}\cdot\vec{r}}{r^3} = \frac{3i}{r}\gamma^5 - \frac{i}{r}\gamma^5 = \frac{2i}{r}\gamma^5$$

$$* \ \left[\Sigma_i,\, \alpha_j p_j\right]\frac{r_i}{r} = 2\varepsilon_{ijk}\alpha_k p_j r_i\left(\frac{1}{r}\right) = 2i\varepsilon_{ijk}\alpha_k\left(r_i p_j - i\delta_{ij}\right)\left(\frac{1}{r}\right) = 2i\alpha_k l_k\left(\frac{1}{r}\right) = \frac{2i}{r}\left(\vec{\Sigma}\vec{l}\right)$$

Grouping these terms together, we derive

$$\left[\vec{\Sigma}\cdot\hat{\vec{r}},\, H\right] = \frac{2i}{r}\gamma^5 + \frac{2i}{r}\gamma^5\left(\vec{\Sigma}\cdot\vec{l}\right) = \frac{2i}{r}\gamma^5\left(\vec{\Sigma}\cdot\vec{l} + 1\right) = \frac{2i}{r}\gamma^5\beta K$$

2) calculation of the commutator $\left[K\left(\vec{\Sigma}\cdot\vec{p}\right),\, H\right]$

$$\left[K\left(\vec{\Sigma}\cdot\vec{p}\right), H\right] = K\left[\vec{\Sigma}\cdot\vec{p},\, H\right] + \left[K, H\right]\left(\vec{\Sigma}\cdot\vec{p}\right) = K\left[\vec{\Sigma}\vec{p},\, \vec{\alpha}\cdot\vec{p} + \beta m + V(r)\right] =$$
$$= K\left[\vec{\Sigma}\vec{p}, \vec{\alpha}\,\vec{p}\right] + mK\left[\vec{\Sigma}\vec{p}, \beta\right] + K\left[\vec{\Sigma}\vec{p}, V(r)\right]$$

Let calculate individual commutators

$$\left[\vec{\Sigma}\vec{p}, \vec{\alpha}\vec{p}\right] - \left[\Sigma_i, \alpha_i\right] p_i p_j = 2i\varepsilon_{ijk}\alpha_k p_i p_j = 0$$

$$\left[\vec{\Sigma}\vec{p},\, \beta\right] = 0$$

$$\left[\Sigma\vec{p},V(r)\right]=\Sigma_i\left[p_i,V(r)\right]=-i\Sigma_i\frac{r_i}{r}V'(r)$$

For the Coulomb potential $V(r)=-\frac{a}{r}\rightarrow V'(r)=\frac{a}{r^2}$ and hence

$$\left[K\left(\vec{\Sigma}\vec{p}\right),H\right]=-i\frac{a}{r^2}K\left(\vec{\Sigma}\cdot\hat{\vec{r}}\right)$$

3) calculation of the commutator $\left[K\gamma^5 f(r),H\right]$

$$\begin{aligned}\left[K\gamma^5 f(r),H\right]&=K\left[\gamma^5 f(r),H\right]+\left[K,H\right]\gamma^5 f(r)=\\&=K\gamma^5\left[f(r),H\right]+K\left[\gamma^5,H\right]f(r)=\\&=K\gamma^5\left[f(r),\vec{\alpha}\,\vec{p}\right]+K2m\gamma^5\beta f(r)=\\&=K\gamma^5 i\vec{\alpha}\hat{\vec{r}}f'(r)+2mK\gamma^5\beta f(r)=\\&=iK\left(\vec{\Sigma}\hat{\vec{r}}\right)f'(r)+2mK\gamma^5\beta f(r)\end{aligned}$$

These relations are used above in calculations of needed relations.

REFERENCES

[1] Sommerfeld A.J.W., *Atombau und Spectrallinien,* Ann.Phys. 50, 1 (1916).

[2] Sukumar C.V., *Supersymmetry and the Dirac Equation for a Central Coulomb field, J.Phys.A: Math.Gen.* 18, L697 (1985)

[3] Johnson M.H., Lippmann B.A. Relativistic Kepler Problem, *Phys. Rev.* 78, 329 (1950).

[4] Katsura H. and Aoki H. Exact Supersymmetry in the Relativistic Hydrogen Atom in General Dimensions - Supercharge and Generalized Johnson-Lippmann operator", *J.Math.Phys.* 47, 032302 (2006).

[5] .Khachidze T.T., Khelashvili A.A.*An* "Accidental" Symmetry Operator for the Dirac Equation in the Coulomb Potential, *Mod. Phys. Lett.* 20, 2277 (2005).

[6] Khachidze T.T., Khelashvili A.A. Hidden Symmetry Operator of the Kepler Problem in Relativistic Quantum Mechanics – from Pauli to Dirac, *Am.J.Phys.* 74, 628 (2006).

[7] de Groot E.H. The Virial Theorem and the Dirac H-Atom, *Am.J.Phys.* 50, 1141 (1982).

[8] Stahlhofen A.A., Algebraic Solutions of Relativistic Coulomb Problems, *Helv. Phys. Acta*, 70, 372 (1997).

Chapter VI

GENERALIZATIONS TO THE RELATIVISTIC DIRAC HAMILTONIAN

"we are all lucky to be living
in an age when we can still
make discoveries"
R.Feynman

VI.1. SUPERSYMMETRY OF THE DIRAC HAMILTONIAN FOR GENERAL CENTRAL POTENTIALS

As we are convinced we have a supersymmetry in the Dirac-Coulomb problem. Moreover if we watch carefully how this symmetry emerges we'll see that SUSY algebra appears immediately as soon as the $K-odd$ operator is constructed.

Therefore it is very interesting to investigate whether there appear some constraints on the potential if we impose supersymmetry on the Dirac Hamiltonian.

So, let us consider the Dirac Hamiltonian again but now for arbitrary central potential $V(r)$:

$$H=\vec{\alpha}\cdot\vec{p}+\beta m+V(r) \qquad \text{(VI.1)}$$

In this form scalar function $V(r)$ is the fourth component of the Lorentz 4-vector. It is easy task to verify, that the K-operator commutes with the Dirac Hamiltonian for arbitrary $V(r)$:

$$[K,H]=0$$

Suppose that there exists some operator Q_1, which anticommutes with K:

$$\{Q_1, K\} = 0$$

Then it is evident that the following operator

$$Q_2 = i\frac{Q_1 K}{\sqrt{K^2}}$$

also anticommutes with K and with Q_1 as well

$$\{Q_2, K\} = 0 \quad \{Q_1, Q_2\} = 0,$$

Moreover

$$Q_1^2 = Q_2^2$$

and let denote it by $\widetilde{H}$. We see that Q_1, Q_2 form $N = 2$ superalgebra (Witten's algebra), where $\widetilde{H}$ plays the role of Witten's Hamiltonian.

In order to have a supersymmetry for the Dirac Hamiltonian (VI.1) we must require that the generators $Q_i \ (i = 1,2)$ commute with H

$$[Q_i, H] = 0 \qquad (i = 1,2) \tag{VI.2}$$

For explicit realization of generators under consideration one has to find $K - odd$ operator(s). One of them is a γ^5 matrix. What else?

Now we can use our theorem and perform a selfconsistent procedure. First of all we can probe the above used combination

$$Q_1 = x_1\left(\vec{\Sigma}\cdot\hat{\vec{r}}\right) + ix_2 K\left(\vec{\Sigma}\cdot\vec{p}\right) + ix_3 K\gamma^5 f(r) \tag{VI.3}$$

Let's calculate

$$[Q_1, H] = \left(\vec{\Sigma}\cdot\hat{\vec{r}}\right)\{x_2 V'(r) - x_3 f'(r)\} + 2i\beta K\gamma^5\left\{\frac{x_1}{r} - mf(r)x_3\right\} = 0$$

Here terms are grouped in a way that we have a diagonal matrix in the first term, while the antidiagonal matrix in the second one. Therefore two equations follow:

$$x_2 V'(r) = x_3 f'(r)$$

$$x_3 m f(r) = \frac{x_1}{r}$$

One find from these equations unambiguously that

$$V(r) = \frac{x_1}{x_2}\frac{1}{mr} \qquad \text{(VI.4)}$$

Hence, in the very general framework we have shown [1] that the *only central potential for which the Dirac Hamiltonian has an additional symmetry (* $N = 2$ *supersymmetry in the above mentioned sense) is a Coulomb potential.*

Meanwhile, the relative signs of coefficients x_1 and x_2 may be arbitrary. Therefore we have symmetry both for attraction and repulsion.

If we take the Coulomb potential in the usual form

$$V(r) = -\frac{a}{r}, \qquad a \equiv Ze^2 \equiv Z\alpha$$

it follows the following relation between parameters involved

$$x_2 = -\frac{1}{ma} x_1$$

In this case our symmetry operator (VI.3) becomes

$$Q_1 = x_1 \left\{ \left(\vec{\Sigma} \cdot \hat{\vec{r}}\right) - \frac{i}{ma} K \left(\vec{\Sigma} \cdot \vec{p}\right) + \frac{i}{mr} K \gamma^5 \right\}$$

Number x_1, as an unessential common factor may be omitted, and after some simple algebra this operator turns to the precise expression of JL operator

$$A = \gamma^5 \left\{ \vec{\alpha} \cdot \hat{\vec{r}} - \frac{i}{ma} K \gamma^5 (H - \beta m) \right\}$$

We know from the previous Section that this operator in the non-relativistic limit reduces to the projection of LRL vector on the electron spin direction. Therefore we see that there is a deep relation between supersymmetry of the Dirac Hamiltonian and the symmetry related to the Laplace - Runge –Lenz vector.

The key feature of supersymmetry is that it unifies the continuous symmetries with the discrete ones of the special kind (like reflections) in a very non–trivial manner. In the transition to relativistic (Dirac) equations we expected that all degeneracies of non–relativistic (Shroedinger) hydrogen atom will be lifted, because the LRL vector is not conserved anymore in this case and only $SO(3)$ symmetry remains. Instead of this owing to underlying hidden symmetry of Coulomb potential the total symmetry rises to $SO(3)\times S(2)$, where $S(2)$ corresponds to the reflection of sign of Dirac's K – operator (discrete transformation).

We have introduced also the so-called Lenz-spin vector $\vec{M}$ (see previous Chapter), which together with the total angular momentum vector $\vec{J}$ generates the $SO(4)$ algebra.

However, the characteristic anticommuting (discrete) relation $\{A,K\}=0$ indicates that the underlying symmetry is not a conventional Lie-symmetry as in non-relativistic case, but generated by a superalgebra.

Nevertheless there still remains residual degeneracy, which is expressed in the suppression of the Lamb shift.

VI.2. Where Is the Harmonic Oscillator?

We are convinced from the previous Section that the Coulomb potential in the Dirac equation is distinguishable among all other central potentials, now from the special point of view – supersymmetry. The same was observed in Classical mechanics when we derived the conserved quantity (the LRL vector) from the classical equations of motion. But in classical mechanics the Kepler problem had a special importance for deriving closed and periodic orbits. Therefore the appearance of vector constant of motion was connected to the closeness of orbits. At the same time let us remember that there was another potential (isotropic harmonic oscillator) which gives also the closed orbits. The corresponding conserved quantity is the symmetric tensor A_{ij}, which are derived by direct inspection.

There appears a natural question: where is a place for this second (i.e. oscillator) potential in the Dirac equation? The commutativity of the Dirac K operator with Hamiltonian, which is a key step for SUSY algebra in our approach, is proved for an arbitrary central potential, is not it?

We'll see below that some caution must be exercised in case of infinitely increasing potentials (class of potentials, to which the oscillator belongs) for the Dirac equation when this potential is a fourth component of the Lorentz four-vector [2]:

$$H=\vec{\alpha}\vec{p}+\beta m+V(r)$$

Let us solve Dirac's stationary equation $H\psi=E\psi$ for this Hamiltonian by standard method, namely, let us present

$$\psi=\begin{pmatrix}\varphi\\ \chi\end{pmatrix}$$

and exclude lower components in obtained equation by the relation

$$\chi = \frac{1}{E+m-V}\vec{\sigma}\vec{p}\varphi$$

Therefore for upper components one gets the following equation

$$\left[\vec{p}^{\,2} - (E-V)^2 + m^2\right]\varphi - i\frac{V'(r)}{E+m-V}\vec{\sigma}\hat{\vec{r}}\varphi = 0 \qquad \text{(VI.5)}$$

It is obvious that if $V(r)$ is increasing potential

$$\lim_{r\to\infty} V(r) = \infty \qquad \text{(VI.6)}$$

then in the configuration space one get the following asymptotic equation

$$\lim_{r\to\infty}\left(\frac{d^2}{dr^2} + V^2(r)\right)\varphi(r) \approx 0$$

As $V(r)$ is increasing, for example, oscillator, this equation have no asymptotically decreasing (normalizable) solutions. It has only oscillating solutions, i.e., this equation has no bound states (Klein paradox, [3]).

The another widespread method for obtaining a second order differential equation is the squaring the Dirac equation i.e. the consideration of the following problem

$$H^2\Psi = E^2\Psi$$

It is evident that after this procedure all the symmetries of the original equation are remained, but it may be that there appear some additional solutions. In our case calculations give

$$\left(\vec{p}^{\,2} + m^2 + V^2 + 2\beta m V + 2V(\vec{\alpha}\cdot\vec{p}) - i(\vec{\alpha}\cdot\hat{\vec{r}})V'(r)\right)\begin{pmatrix}\varphi\\ \chi\end{pmatrix} = E^2\begin{pmatrix}\varphi\\ \chi\end{pmatrix}$$

This has the following asymptotic form for the above class of increasing potentials

$$\left(\vec{p}^{\,2} + V^2\right)\begin{pmatrix}\varphi\\ \chi\end{pmatrix} \approx 0, \qquad r >> 1$$

Therefore the radial asymptotics will be

$$\left(\frac{d^2}{dr^2} - V^2(r)\right)\begin{pmatrix}\varphi(r)\\ \chi(r)\end{pmatrix} \approx 0$$

This equation has normalized solutions with decreasing asymptotics i.e., on the comtrary of previous method, one has solutions corresponding to bound states. So we see that the procedure of such squaring changes the nature of the Dirac equation, there appears new solution, which is not a solution of the original equation.

Thus it is clear that one should handle with caution when infinitely increasing potentials are considered in the Dirac equation. Therefore this classical example (isotropic oscillator potential) has no analogy in the Dirac equation, as compared with the Coulomb potential. These two potentials have different significance for the Dirac equation.

Nevertheless the oscillator problem arises for non–central interactions, switching on the vectorial potentials of the special kinds by gauge covariant manner; for example, considering the following Dirac Hamiltonian

$$H = \vec{\alpha}\left(\vec{p} - \hat{\vec{A}}\right) + \beta m \equiv H_D - \vec{\alpha}\hat{\vec{A}}$$

where $H_D \equiv \vec{\alpha}\vec{p} + \beta m$ and $\hat{\vec{A}}$, generally speaking, can contain some kinds of Dirac's matrices.

The most familiar examples are:

The motion of electron in the uniform magnetic field $\vec{B} = const.$

In this case

$$\hat{\vec{A}} = \vec{A} = \frac{1}{2}B[\vec{n} \times \vec{r}]$$

Here $\vec{n}$ is the unit vector directed along the magnetic field.

By standard manipulations on the Dirac equation, one obtains the Pauli – like equation for the upper component

$$\left(m^2 - E^2\right)\varphi - \left(\vec{p}^2 + \vec{A}^2 + i\vec{\nabla}\vec{A} - 2\vec{A}\vec{p}\right)\varphi -$$
$$-\frac{1}{2}\vec{\sigma}\,\vec{B}\varphi - iB\vec{\sigma}\,\vec{n}(\vec{r}\cdot\vec{p})\varphi = 0$$

It is obvious that if $\left|\vec{A}\right|$ is infinitely increasing, for example, $\left|\vec{A}\right| = \frac{1}{2}Br$, as it is in our case, then $\vec{A}^2$ is dominant term and the problem is reduced to the asymptotic form for the oscillator potential

$$\left(\frac{d^2}{dr^2}-\frac{1}{4}B^2r^2\right)\varphi(r)\approx 0$$

I.e. the Dirac equation in uniform magnetic field has bound states.

If we apply the method of squaring, then we obtain the same equation. So in this example both methods give the same asymptotics.

Last years the following model is fairly considered [4]

$$\hat{\vec{A}} = im\omega\beta\,\vec{r}$$

This is so called "Dirac Oscillator." This model is very interesting because it reduces to the oscillator problem and exhibits a supersymmetry.

VI.3. Relativistic Quantum Mechanics of Dirac Oscillator

Moshinsky and Szczepaniak [4] introduced very interesting potential in the Dirac equation. They considered a system whose Dirac Hamiltonian is linear in both $\vec{p}$ and $\vec{r}$. Consequently, its non–relativistic approximation is quadratic in $\vec{r}$ as in the case of the harmonic oscillator. This property is in fact the origin of the name "Dirac oscillator". The system is then, excepting for a strong spin–orbit coupling term, the square root of the harmonic oscillator in the same sense as the Dirac equation is the square root of the Klein – Gordon equation.

The Hamiltonian of the system is obtained by performing the following non – minimal substitution in the free Dirac particle's Hamiltonian

$$\vec{p} \rightarrow \vec{p} - im\omega\beta\vec{r}$$

where m is the mass of the particle described and ω is the oscillator frequency, β is the usual Dirac matrix.

The Dirac equation for the system is

$$i\frac{\partial\Psi}{\partial t} = H\Psi = \left[\vec{\alpha}\cdot\left(\vec{p} - im\omega\beta\,\vec{r}\right) + m\beta\right]\psi \tag{VI.7}$$

It is obvious that despite presence of non–minimal term, the resulting Hamiltonian still remains hermitian. The interaction introduced by this procedure has been shown to correspond to an anomalous magnetic interaction. Moreover it is straightforward to show that the system conserves the total angular momentum [4] $\vec{j} = \vec{L} + \frac{1}{2}\vec{\Sigma}$. Also the Dirac's K-operator commutes with this Hamiltonian.

If we select a frame dependent vector [5]

$$U^{\mu} = m\omega\left(1,\vec{0}\right) \tag{VI.8}$$

then the interaction term can be put in the form

$$im\omega\vec{\alpha}\vec{r} = \sigma^{\mu\nu} x_{\mu} U_{\nu}$$

And consequently, the equation of motion can be written in the manifestly covariant manner

$$\left[\gamma^{\mu} P_{\mu} - m + \frac{ke}{4m}\sigma^{\mu\nu} F_{\mu\nu}\right]\Psi = 0 \tag{VI.9}$$

with $F_{\mu\nu} = \partial_{\mu} A_{\nu} - \partial_{\nu} A_{\mu}$ and $k = \dfrac{2m}{e}$ is the anomalous magnetic moment of the Dirac oscillator. The electromagnetic potential is

$$A_{\mu} = \frac{1}{2}\left[x_{\mu}(U \cdot x) - \frac{1}{2}x^{2} U_{\mu}\right]$$

where U^{μ} is interpreted as the four–velocity of a point–particle attached to the origin of the laboratory frame, which is the proper frame of uniformly charged dielectric sphere. Using the gauge invariance of the electromagnetic interaction and selecting the gauge as $\lambda = \dfrac{1}{4}x^{2}(U \cdot x)$, the vector potential may be rewritten as [5]

$$A_{\mu}^{\pm} = A_{\mu} \pm \partial_{\mu}\lambda = \begin{cases} \dfrac{1}{2}(U \cdot x)x_{\mu} \\ -\dfrac{1}{2}x^{2} U_{\mu} \end{cases}$$

For either selection the electromagnetic field takes the form $\vec{E} = -m\omega\vec{r}$ and $\vec{B} = 0$.

In this way it was shown [6] that the interaction occurs as if it was produced by an infinite sphere carrying a uniform charge–density distribution, resulting in a linearly growing electric field.

The interaction term is interesting in its own right, but also has application in QCD as an effective chromoelectric field.

From the structure of the Hamiltonian it follows that parity is a good quantum number. Moreover $j = l \pm \frac{1}{2}$, where j is the total angular momentum quantum number and l is the orbital one. The parity of the energy eigenstate is given by $(-1)^l$, we therefore define

$$\varepsilon = \begin{cases} +1, & if\ parity\ is\ (-1)^{j+\frac{1}{2}} \\ -1, & if\ parity\ is\ (-1)^{j-\frac{1}{2}} \end{cases} \quad \text{(VI.10)}$$

In both cases we have $l = j + \dfrac{\varepsilon}{2}$

Now the solution of the Dirac equation may be written as

$$\Psi(x) = \Psi(\vec{r}, t) = \frac{1}{r}\begin{Bmatrix} F(r)\mathrm{Y}_l(\theta,\varphi) \\ iG(r)\mathrm{Y}_{l'}(\theta,\varphi) \end{Bmatrix} \exp(-iEt) \quad \text{(VI.11)}$$

where Y_l and $\mathrm{Y}_{l'}$ are spinor spherical harmonics of order l and l', respectively. Here $l' = j - \frac{\varepsilon}{2}$ as $\mathrm{Y}_{l'}$ must be of opposite parity to Y_l.

The radial components F and G are solutions to the following coupled differential equations

$$\left\{ -\frac{d}{dr} + \frac{1}{r}\left[\varepsilon\left(j + \frac{1}{2} \right) + m\omega r^2 \right] \right\} G(r) = (E - m)F(r) \quad \text{(VI.12)}$$

and

$$\left\{ +\frac{d}{dr} + \frac{1}{r}\left[\varepsilon\left(j + \frac{1}{2} \right) + m\omega r^2 \right] \right\} F(r) = (E + m)G(r) \quad \text{(VI.13)}$$

Excluding $G(r)$ from this system the following second order differential equation follows for the upper (large) component

$$(E^2 - m^2)F(r) = \left\{ -\frac{d^2}{dr^2} + \frac{\left(j + \frac{1}{2}\right)\left(j + \frac{1}{2} + \varepsilon\right)}{r^2} + m^2\omega^2 r^2 + m\omega\left[\varepsilon(2j+1) - 1\right] \right\} F(r) \quad \text{(VI.14)}$$

Exactly this equation follows by squaring of our Hamiltonian

$$H^2 = \left[\vec{\alpha}\cdot(\vec{p} - im\omega\beta\vec{r}) + m\beta\right]^2 = p^2 + m^2\omega^2 r^2 + \left(2\vec{\sigma}\cdot\vec{L} - 3\right)m\omega\beta + m^2, \quad \text{(VI.15)}$$

because

$$\vec{\sigma} \cdot \vec{L} = -\frac{1}{2}\varepsilon(2j+1)\beta - 1$$

So in this case both methods of obtaining the second order differential equations give the same results.

We see that the square of the Hamiltonian for the Dirac Oscillator becomes essentially a harmonic oscillator plus constant terms. Moreover the squared Hamiltonian is composed only of even operators (β-matrix) then a closed form for the Foldy – Wouthuysen transformation [7] can be found.

We mention also that this system has some remarkable symmetry properties. First, as it conserves angular momentum it must be $O(3)$ invariant, but in fact that it admits the larger dynamical symmetry algebra $SO(3.1) \otimes SO(4)$, thus producing a more degenerate spectrum than expected from $O(3)$ considerations only (see below). Second, the existence of a closed Foldy-Wouthyusen transformation implies that the positive and negative energy solutions never mix independently of the intensity of the interaction.

This problem can be exactly solved with no more effort than a non– relativistic harmonic oscillator; it can even be solved using purely algebraic techniques.

Now we mention that the standard methods give the solution to the above system of equations (VI.12-13) in the following form:

$$F_{n,l}(r) = A\left[(m\omega)^{1/2} r\right]^{l+1} \exp\left(-\frac{m\omega r^2}{2}\right) L_n^{l+1/2}\left(m\omega r^2\right) \tag{VI.16}$$

and

$$G_{n,l}(r) = A\left[(m\omega)^{1/2} r\right]^{l'+1} \exp\left(-\frac{m\omega r^2}{2}\right) L_n^{l'+1/2}\left(m\omega r^2\right) \tag{VI.17}$$

where $L_\alpha^\beta(x)$ is an associated Laguerre Polynomial, and $N = 2n + l$, $n = 0,1,2,...$, is the principal quantum number. A is the normalization constant.

The energy spectrum can be obtained from

$$E^2 - m^2 = m\omega[2(N+1) + \varepsilon(2j+1)] \tag{VI.18}$$

for the positive – energy state, and from

$$E^2 - m^2 = m\omega[2(N+2) + \varepsilon(2j+1)] \tag{VI.19}$$

for the negative energy states.

It is important to point out here that the complete energy spectrum is given by both of equations above and be written as

$$E_{Nlj}^2 - m^2 = m^2 \begin{cases} \omega[2(N-j)+1], & if \ \ l = j - \tfrac{1}{2} \\ \omega[2(N+j)+3], & if \ \ l = j + \tfrac{1}{2} \end{cases} \tag{VI.20}$$

We clearly see the existence of an infinite degeneracy if $l = j - \tfrac{1}{2}$ as the (N, j) pair has the same energy as all the pairs

$$(N \pm 1,\ j \pm 1),\ \ (N \pm 2,\ j \pm 2), \ldots$$

and the series is not bounded from above.

On the other hand if $l = j + \tfrac{1}{2}$ there is the same energy for all the pairs

$$(N \pm 1,\ j \mp 1),\ \ (N \pm 2,\ j \mp 2), \ldots$$

and the series cuts off both at $j = \tfrac{1}{2}$ and $N = 0$, so there is a finite degeneracy.

Moshinsky and Quesne have shown that the symmetry Lie algebra responsible for the infinite accidental degeneracy in the Dirac oscillator i.e. when $l = j - \tfrac{1}{2}$, is the Lorentz algebra $O(3,1)$, while in the finite case $l = j + \tfrac{1}{2}$ the symmetry Lie algebra is the orthogonal one $O(4)$.

Let us address now the problem of the hidden supersymmetry.

If we set

$$\Phi = \begin{pmatrix} F \\ G \end{pmatrix}$$

then we can construct an equivalent supersymmetric Hamiltonian H_S acting only on the radial part of the eigenfunctions,

$$H_S \Phi = (E^2 - m^2)\Phi \tag{VI.21}$$

Here we define

$$H_S = \{Q, Q^+\} = -\frac{d^2}{dr^2} + \frac{\left[\varepsilon\left(j+\frac{1}{2}\right)+r^2\right]^2}{r^2} - \left(1 - \frac{\varepsilon\left(j+\frac{1}{2}\right)}{r^2}\right)\sigma_3 \qquad \text{(VI.22)}$$

where Q and Q^+ are fermionic operators, the generators of SUSY transformations, defined as

$$Q = (p + iW')\sigma^+ \text{ and } Q^+ = (p - iW')\sigma^-$$

with

$$p = \frac{1}{i}\frac{d}{dr}, \quad \sigma^\pm = \sigma_1 \pm i\sigma_2$$

And the superpotential $W(r)$ is given by

$$W(r) = \varepsilon\left(j + \frac{1}{2}\right)\ln r + \frac{1}{2}m\omega^2 r^2 \qquad \text{(VI.23)}$$

Therefore, we see that the large radial component $F(r)$ is a solution of the fermionic sector of H_S and is the supersymmetric partner of $G(r)$.

In summary, it was shown that the Dirac oscillator is a system exactly solvable, it conserves angular momentum and it is such that the positive and negative energy states never mix, is intimately related with a hidden supersymmetry produced by the "anomalous momentum" interaction $im\omega\beta r$.

VI.4. The Lorentz - Scalar Potential in the Dirac Equation

Another important example of a non–minimal coupling in the Dirac equation is a Lorentz–scalar potential. This potential together with the 4th component of Lorentz–vector completes most general central interaction in the Dirac equation.

Let us consider the full Hamiltonian

$$H = \vec{\alpha} \cdot \vec{p} + \beta m + V(r) + \beta S(r) \qquad \text{(VI.24)}$$

Here $S(r)$ is a Lorentz–scalar, while $V(r)$ is only $O(3)$ scalar, but a 4th component of Lorentz–vector, as in sec.VI.1.

We saw that in case of $V(r)=V_{Coulomb}$ the Dirac Hamiltonian commutes with the JL operator. Now this Hamiltonian still commutes with Dirac's K-operator, but does not commute with the above JL operator even for Coulomb potential.

It is evident that non–relativistic quantum mechanics is indifferent as regards to the Lorentz transformation (variance) properties of potential, therefore it is expected that in case of scalar potential the description of hidden symmetry should also be possible. In other words, the JL operator must be generalized.

For this purpose we make use of our method based on theorem about $K-odd$ structures as compared with the previous case, particulary in the part of additional $\hat{O}$ factors.

Therefore we probe the following operator

$$Q_1 \equiv X = x_1\left(\vec{\Sigma}\cdot\hat{\vec{r}}\right)+x_1'\left(\vec{\Sigma}\cdot\hat{\vec{r}}\right)H + ix_2K\left(\vec{\Sigma}\cdot\vec{p}\right)+ix_3K\gamma_5 f_1(r)+ix_3'K\gamma_5\beta f_2(r)$$

We have included $\hat{O}=H$ in the first structure and at the same time the matrix $\hat{O}=\beta$ in the third one. Both of them commute with K. It is a minimal extension of the previous picture, when only the first order structures in $\hat{\vec{r}}$ and $\vec{p}$ participate. For turning to the previous case one must take $x_1' = x_3' = 0$ and $S(r)=0$.

Calculations of relevant commutators give

$$\begin{aligned}
[X,H] = {} & \gamma^5\beta K\left\{\frac{2ix_1}{r} - 2ix_3(m+S)f_1(r)+\frac{2ix_1'}{r}V(r)\right\}+ \\
& + K\left(\vec{\Sigma}\cdot\hat{\vec{r}}\right)\left\{x_2V'(r)-x_3f_1'(r)\right\}+ \\
& + K\beta\left(\vec{\Sigma}\cdot\hat{\vec{r}}\right)\left\{x_2S'(r)-x_3'f_2'(r)\right\}+ \\
& + \gamma^5K\left\{\frac{2ix_1'(m+S)}{r} - 2ix_3'(m+S)f_2(r)\right\}+ \\
& + \beta K\left\{\frac{2ix_1'}{r} - 2ix_3'f_2(r)\right\}\left(\vec{\Sigma}\cdot\vec{p}\right)
\end{aligned}$$

Equating this expression to zero, we derive matrix equation, then after passing to 2×2 representation we must equate to zero separately the coefficients standing in fronts of diagonal and antidiagonal elements.

In this way it follows equations:

1) from diagonal structures ($K\left(\vec{\Sigma}\cdot\hat{\vec{r}}\right)$, $K\beta\left(\vec{\Sigma}\cdot\hat{\vec{r}}\right)$, $\beta K\left(\vec{\Sigma}\cdot\vec{p}\right)$):

$$x_2 V'(r) - x_3 f_1'(r) = 0$$

$$\frac{x_1}{r} - x_2(m+S)V(r) + \frac{x_1'}{r}V(r) = 0$$

$$x_2 S'(r) - x_3' f_2'(r) = 0$$

$$\frac{x_1'}{r} - x_3' f_2(r) = 0$$

2) from antidiagonal structures $(\gamma^5 K, \ \gamma^5 \beta K)$:

$$\frac{x_1}{r} - x_3(m+S)f_1(r) + \frac{x_1'}{r}V(r) = 0$$

$$\frac{x_1'}{r}(m+S) - x_3'(m+S)f_2(r) = 0$$

Integrating the first and third equations of 1) for vanishing boundary conditions at infinity, we obtain

$$f_1(r) = \frac{x_2}{x_3}V(r), \qquad f_2(r) = \frac{x_2}{x_3'}S(r)$$

and taking into account the last equations from 1) and 2), we have

$$f_2(r) = \frac{x_1'}{x_3' r}$$

Therefore, according to previous relation for $S(r)$, we obtain finally

$$S(r) = \frac{x_1'}{x_2 r} \tag{VI.25}$$

So, the scalar potential must be Coulombic.

Inserting now f_1 into the first equation of 2) and solving for $V(r)$, one derives

$$V(r) = \frac{x_1}{r}\frac{1}{x_2(m+S) - \frac{x_1'}{r}},$$

At last, using here derived expression for $S(r)$, we find

$$V(r) = \frac{x_1}{x_2 m r} \tag{VI.26}$$

Therefore we make sure that the $N = 2$ supersymmetry in the above described content is symmetry of the Dirac Hamiltonian only for Coulomb potential (for any general combination of Lorentz–scalar and 4th component of a Lorentz–vector).

Now if we take into account above obtained relations and use them into the general expression for X , one can reduce it to more compact form [8]

$$\mathrm{X} = \left(\vec{\Sigma}\cdot\hat{\vec{r}}\right)\left(ma_V + Ha_S\right) - iK\gamma_5\left(H - \beta m\right) \tag{VI.27}$$

where the following notations are used:

$$a_V = -\frac{x_1}{x_2 m}, \qquad a_S = -\frac{x_1'}{x_2}$$

Here $a_i - s$ are the constants of corresponding Coulomb potentials

$$V(r) = -\frac{a_V}{r}, \quad S(r) = -\frac{a_S}{r},$$

It is remarkable that the above result is derived by general 3–dimensional approach without referring to radial equation, as in earlier investigation [9]. Therefore it is more systematic and transparent.

VI.5. Algebraic Derivation of the Spectrum of the Dirac Hamiltonian for an Arbitrary Combination of the Lorentz-Scalar and Lorentz-Vector Coulomb Potential

We have demonstrated above the efficiency of algebraic methods. We elucidated that the Witten's $N = 2$ superalgebra rises immediately as soon as an operator anticommuting with K is constructed. In our case it is only sufficient to introduce the supersymmetric generators as follows

$$Q_1 = \mathrm{X}, \qquad Q_2 = i\frac{\mathrm{X}K}{|K|}$$

Then the anticommutativity $\{X, K\} = 0$, yields

$$\{Q_1, Q_2\} = 0, \quad Q_1^2 = Q_2^2 \equiv \widetilde{H}$$

Therefore, we are faced agan with the Witten's Hamiltonian

Now we want to obtain spectrum of the Dirac Hamiltonian pure algebraically, without any referring on equations of motion. Our method is based on Witten's superalgebra, established above.

To explore this algebra, one defines a SUSY ground state $|0\rangle$:

$$\widetilde{H}|0\rangle = \mathrm{X}^2|0\rangle = 0 \quad \to \mathrm{X}|0\rangle = 0 \tag{VI.28}$$

Because X^2 is a square of Hermitian operator, it has a positive definite spectrum and one is competent to take zero this operator itself in ground state. By this requirement we'll obtain Hamiltonian in this ground state and, correspondingly, ground state energy. After that by well known ladder procedure one can construct the energies of all excited levels.

Let equate $\mathrm{X} = 0$ and solve H from Eq. (VI.27)):

$$H = m\left[\left(\vec{\alpha}\cdot\hat{\vec{r}}\right)a_S + iK\right]^{-1}\left[iK\beta - a_V\left(\vec{\alpha}\cdot\hat{\vec{r}}\right)\right] = \frac{m}{\kappa^2 + a_S^2}N$$

where

$$\begin{aligned} N \equiv & \left[\left(\vec{\alpha}\cdot\hat{\vec{r}}\right)a_S - iK\right]\left[iK\beta - a_V\left(\vec{\alpha}\cdot\hat{\vec{r}}\right)\right] = \\ & = -a_S a_V + K\left[K\beta + ia_V\left(\vec{\alpha}\cdot\hat{\vec{r}}\right)\right] - ia_S K\beta\left(\vec{\alpha}\cdot\hat{\vec{r}}\right) \end{aligned} \tag{VI.29}$$

Now we try to diagonalize this operator using Foldy-Wouthuysen [7] like transformation. Because the second and third terms do not commute with each others we need several (at least two) such transformations.

We choose the first transformation in the following manner

$$\exp(iS_1) = \exp\left(-\frac{1}{2}\beta\left(\vec{\alpha}\cdot\hat{\vec{r}}\right)w_1\right) \tag{VI.30}$$

It is evident that

$$\begin{aligned} &\exp(iS_1)\left(\vec{\alpha}\cdot\hat{\vec{r}}\right)\exp(-iS_1) = \exp(2iS_1)\left(\vec{\alpha}\cdot\hat{\vec{r}}\right) \\ &\exp(iS_1)\beta\exp(-iS_1) = \exp(2iS_1)\beta \end{aligned}$$

Moreover

$$\exp(iS_1)K\exp(-iS_1) = K, \qquad \exp(iS_1)\beta K\exp(-iS_1) = \exp(2iS_1)\beta K$$

and

$$\exp(iS_1)\beta(\vec{\alpha}\cdot\hat{\vec{r}})\exp(-iS_1)=\beta(\vec{\alpha}\cdot\hat{\vec{r}})$$

Therefore the first transformation acts as

$$N'\equiv\exp(iS_1)N\exp(-iS_1)=-a_S a_V+K\exp(2iS_1)\left[K\beta+ia_V(\vec{\alpha}\cdot\hat{\vec{r}})\right]-ia_S K\beta(\vec{\alpha}\cdot\hat{\vec{r}}) \quad \text{(VI.31)}$$

But

$$\exp(2iS_1)=chw_1+i\beta(\vec{\alpha}\cdot\hat{\vec{r}})shw_1$$

Make use of this relation, we have

$$\begin{aligned}&\exp(2iS_1)\left[K\beta+ia_V(\vec{\alpha}\cdot\hat{\vec{r}})\right]=\\&=\beta\left[Kchw_1+a_V shw_1\right]+K(\vec{\alpha}\cdot\hat{\vec{r}})\left[ia_V chw_1+iKshw_1\right]\end{aligned} \quad \text{(VI.32)}$$

Now in order to get rid of non-diagonal $(\vec{\alpha}\cdot\hat{\vec{r}})$ terms, we must choose

$$thw_1=-\frac{a_V}{K} \quad \text{(VI.33)}$$

Using simple trigonometric relations we arrive at

$$\exp(2iS_1)\left[K\beta+ia_V(\vec{\alpha}\cdot\hat{\vec{r}})\right]=K^{-1}\beta\sqrt{\kappa^2-a_V^2} \quad \text{(VI.34)}$$

Let us perform the second F.-W. transformation

$$N''=\exp(iS_2)N'\exp(-iS_2),$$

where

$$S_2=-\frac{1}{2}(\vec{\alpha}\cdot\hat{\vec{r}})w_2 \quad \text{(VI.35)}$$

Now

$$\begin{aligned}&\exp(iS_2)K\beta\exp(-iS_2)=\exp(2iS_2)K\beta\\&\exp(iS_2)K\beta(\vec{\alpha}\cdot\hat{\vec{r}})\exp(-iS_2)=\exp(2iS_2)K\beta(\vec{\alpha}\cdot\hat{\vec{r}})\\&\qquad\exp(2iS_2)=\cos w_2-i(\vec{\alpha}\cdot\hat{\vec{r}})\sin w_2\end{aligned}$$

Therefore

$$N'' = -a_S a_V + K\sqrt{\kappa^2 - a_V^2}\exp(2iS_2)\beta - ia_S \exp(2iS_2)K\beta(\vec{\alpha}\cdot\hat{\vec{r}}) =$$
$$= -a_S a_V + K\sqrt{\kappa^2 - a_V^2}\beta\cos w_2 + iK\sqrt{\kappa^2 - a_V^2}\beta(\vec{\alpha}\cdot\hat{\vec{r}})\sin w_2 -$$
$$- ia_S K\beta(\vec{\alpha}\cdot\hat{\vec{r}})\cos w_2 + a_S K\beta\sin w_2 \quad \text{(VI.36)}$$

Requiring absence of $(\vec{\alpha}\cdot\hat{\vec{r}})$ terms we have

$$tgw_2 = \frac{a_S}{\sqrt{\kappa^2 - a_V^2}} \quad \text{(VI.37)}$$

Therefore

$$N'' = -a_S a_V + K\beta\sqrt{\kappa^2 - a_V^2 + a_S^2} \quad \text{(VI.38)}$$

and finally

$$H = \frac{m}{\kappa^2 + a_S^2}\left\{-a_S a_V + K\sqrt{\kappa^2 - a_V^2 + a_S^2}\,\beta\right\} \quad \text{(VI.39)}$$

For eigenvalues in ground state we have

$$E_0 = \frac{m}{\kappa^2 + a_S^2}\left\{-a_S a_V \pm \kappa\sqrt{\kappa^2 - a_V^2 + a_S^2}\right\} \quad \text{(VI.40)}$$

Now let us remember the result obtained by explicit solution of the Dirac equation for this case [10]

$$E = m\left\{\frac{-a_S a_V}{a_V^2 + (n - |k| + \gamma)^2} \pm \sqrt{\left(\frac{a_S a_V}{a_V^2 + (n - |\kappa| + \gamma)^2}\right)^2 + \frac{(n - |\kappa| + \gamma)^2 - a_S^2}{a_V^2 + (n - |\kappa| + \gamma)^2}}\right\} \quad \text{(VI.41)}$$

where

$$\gamma^2 = \kappa^2 - a_V^2 + a_S^2 \quad (51)$$

In the ground state $n = 1,\quad j = 1/2 \;\rightarrow\; |\kappa| = j + 1/2 = 1$, there remains

$$E_0 = m\left\{\frac{-a_S a_V}{a_V^2+\gamma^2} \pm \sqrt{\left(\frac{a_S a_V}{a_V^2+\gamma^2}\right)^2 + \frac{\gamma^2 - a_S^2}{a_V^2+\gamma^2}}\right\} \qquad \text{(VI.42)}$$

this after obvious manipulations reduces to our above derived expression (VI.40).

Therefore by only algebraic methods we have obtained the correct expression for ground state energy.

For obtaining of total spectrum it is sufficient now to use the Witten's algebra. Following to the ordinary ladder procedure, this consists in change (for our case):

$$\gamma \quad \to \quad \gamma + n - |\kappa|$$

Making use of this, the correct expression for total energy spectrum, eq. (VI.41) follows.

Comments

Supersymmetry requirement appears to be a very strong constraint in the framework of the Dirac Hamiltonian. While the supercharge operator, commuting with the Dirac Hamiltonian in case of pure vector component only is intimately related to the LRL vector, but unlike the latter one relativistic supercharge participates in transformations of spin degrees of freedom. In passing to non-relativistic physics, information concerning to spin–degrees of freedom disappears and hence LRL vector as a generator of algebra does not transform anything and symmetry becomes hidden as a relic of relativistic quantum mechanics.

A standard way for motivating the Coulomb potential is one photon exchange in the framework of quantum electrodynamics. Interestingly enough that the uniqueness of Coulomb potential can be proved requiring the supersymmetry ($N = 2$ in the presented above sense) of the Dirac Hamiltonian as a theoretical ground for it. It has been known for a long time that the Dirac Hamiltonian with the Coulomb potential exhibits $N = 2$ supersymmetry. Things turn out to be more dramatic however; supersymmetry requirement leads us uniquely to the Coulomb potential in the framework of the Dirac Hamiltonian for general central interactions.

Moreover the spectrum of the Dirac equation for general Coulomb potential (vector and scalar combination) is derived algebraically, without consideration or solution of radial differential equations of motion. So it is proved that this problem is totally integrable.

REFERENCES

[1] Khachidze T.T., Khelashvili A.A., *Supercharge Operator for Hidden Symmetry in the Dirac Equation.*, In Proceedings of CICHEP II. Vol.881, 279. (Ed. S.Khalil), Melville, New York, 2007.

[2] Khachidze T.T., Khelashvili A.A.Nadareishvili T.P., Dirac Equation and its Squared Form. Bull. Georg. *Acad. Sci.*, 174, 61 (2006).

[3] Bjorken J.D., Drell S.D., *Relativistic Quantum Mechanics.* New York, McGraw-Hill, 1964.

[4] Moshinsky M., Szczepaniak A., The Dirac Oscillator. *J.Phys. A: Math.Gen.* 22,L817 (1989).

[5] Benitez J. et al., Solution and Hidden Supersymmetry of a Dirac Oscillator. *Phys. Rev. Lett.* 64, 1643 (1990).

[6] Martnez-y-Romero R.P. et al., *Relativistic Quantum Mechanics of a Dirac Oscillator.* arXiv: quant-p/9908069 (1999).

[7] Foldy L.and Wouthuysen S. On theDirac Spin-1/2 Particle and Its Non-Relativistic Limit. *Phys. Rev.* 72, 29 (1950).

[8] Khachidze T.T.and Khelashvili A.A., Algebraic Derivation of the Spectrum of the Dirac Hamiltonian for an Arbitrary Combination of the Lorentz-scalar and Lorentz-vector Coulomb Potentials, *Ukr. Fiz. Journ.* 52, 421 (2007).

[9] Leviatan A. Supersymmetry Pattern in the Pseudospin Spin and Coulomb Limits of the Dirac Equation with Scalar and Vector Potenials. *Phys. Rev. Lett.* 92, 20201 (2004).

[10] Greiner W., Muller B. and Rafelski J. *Quantum Electrodynamics of Strong Fields.* Springer-Verlag, 1985.

Chapter VII

SOME RECENT DEVELOPMENTS

VII.1 HIDDEN SUPERSYMMETRY OF THE DIRAC-COULOMB PROBLEM AND THE BIEDENHARN APPOACH

In the previous Chapter the three dimensional Dirac-Coulomb problem was considered. But our consideration would not be perfect without discussing of the radial equations as well especially from SUSY point of view.

The exact solvability of the Dirac-Coulomb problem can be explained with the help of SUSY quantum mechanics. SUSY affords an elegant interpretation of the level degeneracies of a Dirac electron both in a constant magnetic field and also in the hydrogen atom.

In the Dirac-Coulomb system, described by Sukumar (see, Sec.V, Ref.[2]) constructon centres in the coupled first-order radial equations. There is a different approach based on squared form of the Dirac equation [1-3], the associated second-order radial equation may be cast in scalar form, which is very close to non-relativistic Schroedinger equation and particularly emenable to the standard application of supersymmetric quantum mechanics.

As we already know an application of SUSY ideas for the non-relativistic hydrogen atom generates radial operators for adjacent l values (but the same potential) as supersymmetric partners. Hence this supersymmetry implies the l independence of the eigenvalues traditionally associated with $O(4)$ symmetry of this problem.

The factorization technique has been applied in the case of relativistic hydrogen atom both to the second-order radial equaton and to the coupled first-order radial equations. For any choice of total momentum quantum number j the energy level structure of the Dirac atom, apart from an overall shift in energy, has precisely the standard form of a supersymmetric spectrum: a non-degenerate ground state, and doubly degenerate excited states. This indicates strongly that this degeneracy may be attributed to supersymmetry on essentially the same manner as for non-relativistic hydrogen atom. It appears as if this supersymmetry should result from a trivial application of the general formulation to the second-order radial equation for the relativistic hydrogen atom.

However, some differences of principle between the solutions of the relativistic and non-relativistic second-order equation complicate the demonstration. First of all, we no longer have $O(4)$ symmetry, the analogue of l in the radial equation is not an integer and the parameters in the radial equation are level-dependent. Moreover there appears new non-hermitian operator, wich characterizes the energy levels and plays the same role as orbital momentum in non-relativistic approach, but now it has “irrational” eigenvalues.

As we know the Dirac-Coulomb problem involves the Hamiltonian

$$H = \vec{\alpha}\vec{p} + \beta m - \frac{Ze^2}{r}$$

We propose to solve the eigenvalue equation

$$H\psi = E\psi \tag{VII.1}$$

in the following way:

Rewrite eq. (VII.1) in the form

$$O_+\psi = 0 \tag{VII.2}$$

where

$$O_+ \equiv \beta(H - E) = \beta(\vec{\alpha}\vec{p}) - \beta\left(E + \frac{Ze^2}{r}\right) + m \tag{VII.3}$$

2. Introduce the operator O_- given by

$$O_- \equiv \beta(\vec{\alpha}\vec{p}) - \beta\left(E + \frac{Ze^2}{r}\right) - m \tag{VII.4}$$

It is evident that these two operators are commuting

$$[O_+, O_-] = 0$$

3. Consider the product of them O_-O_+ which is a second order differential operator, and assume that the function Φ is a solution of the equation

$$O_-O_+\Phi = 0 \tag{VII.5}$$

We then see that the desired solution to the Dirac equation is given by the projection operator O_-:

$$\psi = O_-\Phi \tag{VII.6}$$

and conversely, every solution to the Dirac equation satisties eq. (VII.5).

It is evident from the explicit form of the JL operator that it does not commute with this operator product because of its dependence on the sign in front of mass. So the JL operator is no longer symmetry of squared equation.

Consider now the second order operator given in (VII.5)

$$O_-O_+ = \left(E + \frac{Ze^2}{r}\right)^2 - \vec{p}^{\,2} - m^2 - i\left(\vec{\alpha}\hat{\vec{r}}\right)\frac{Ze^2}{r^2} \qquad \text{(VII.7)}$$

(In general, if instead of Coulomb potential an arbitrary $V(r)$ is considered, the last term in this relation would be $-i\left(\vec{\alpha}\hat{\vec{r}}\right)V'(r)$, but only for Coulomb potential arises r^{-2} term and corresponding symmetries appear only for this case!

(VII.5) is Krammer's equation. We note that O_- is not the Hermitian adjoint of O_+. The operator O_-O_+ is therefore not Hermitian.

btained operator (VII.7) differs from the Klein-Gordon operator for the Kepler problem solely in the occurrence of the last term.

If we remember the Dirac operator $K = \beta\left(\vec{\Sigma}\vec{l} + 1\right)$, we can rewrite the obtained second-order differential equation as

$$\left(\frac{1}{r^2}\frac{d}{dr}r^2\frac{d}{dr} - \frac{K^2 - (\alpha Z)^2}{r^2} + \frac{2\alpha ZE}{r} - \frac{1}{r^2}\left(K + i\alpha Z\left(\vec{\alpha}\hat{\vec{r}}\right) - k^2\right)\Phi = 0\right) \qquad \text{(VII.8)}$$

At the first glance the radial equation looks quite complicated but it can be simplified greatly if we introduce so called Biedenharn (Temple) operator [1-4], defined as

$$\Gamma \equiv \beta K + i\alpha Z\left(\vec{\alpha}\hat{\vec{r}}\right) \qquad \text{(VII.9)}$$

This operator Γ was introduced by Martin and Glauber [4] and a radial supersymmetry induced by Γ-operator was studied by Dahl and Jorgensen [5]. This operator was introduced by Temple [6].

Consideration of this operator allows us to rewrite equation in a form reminiscent of the non-relativistic Coulomb problem

$$\left[-\left(\frac{d}{dr} + \frac{1}{r}\right)^2 + \frac{\Gamma(\Gamma+1)}{r^2} - \frac{2\alpha ZE}{r} + m^2 - E^2\right]\psi = 0 \qquad \text{(VII.10)}$$

The operator Γ plays here a role of the angular momentum. Because the two terms in Γ anticommute, we have a remarkable relation

$$\Gamma^2 = K^2 - (\alpha Z)^2 = \vec{J}^2 + \frac{1}{4} - (\alpha Z)^2 \qquad \text{(VII.11)}$$

So that the eigenvalues of Γ are

$$\gamma = \pm\sqrt{\kappa^2 - (\alpha Z)^2} = \pm\sqrt{(j+1/2)^2 - (\alpha Z)^2}, \qquad \operatorname{sgn}(\gamma) = \operatorname{sgn}(\kappa)$$

and

$$\Gamma(\Gamma+1) = l(\gamma)(l(\gamma)+1), \qquad \textit{with} \quad l(\gamma) = |\gamma| + 1/2(\operatorname{sgn}(\gamma) - 1) \qquad \text{(VII.12)}$$

I.e. the "angular momentum" $l(\gamma)$ is *irrational.*

Therefore we see that the result is formally identical to the radial differential equation for the non-relativistic Kepler problem except for a different meaning ascribed to the numerical parameters. That is:

- In place of non-relativistic wave number $k_{NR} = (2m|E|)^{1/2}$, we have the relativistic wave number $k = k_{rel} = (m^2 - E^2)^{1/2}$.
- In place of the non-relativistic Sommerfeld parameter $\eta_{NR} = \frac{m\alpha Z}{k_{NR}}$, we have the relativistic Sommerfeld parameter $\eta_{rel} = \frac{\alpha ZE}{k_{rel}}$.

the most stricking "parameter change" is that in place of the non-relativistic centrifugal term $\frac{1}{r^2}L(L+1)$ we have the relativistic term $\frac{1}{r^2}\Gamma(\Gamma+1)$. If we bring the operator Γ to the diagonal form, it involves the replacement of the integer $|\kappa|$ by the nonintegral $\gamma|\kappa| = \left(\kappa^2 - (\alpha Z)^2\right)^{1/2}$, which defines the relativistic eccentricity in the Sommerfeld frame.

The transformation that brings Γ to diagonal form $\Gamma \to \gamma\kappa$, $\gamma\kappa = \pm\left(\kappa^2 - (\alpha Z)^2\right)^{1/2}$ is given by the operator [1]:

$$S = \exp\left(-\frac{1}{2}i(\vec{\alpha}\hat{\vec{r}})\tanh^{-1}\left(\frac{\alpha Z}{K}\right)\right) \qquad \text{(VII.13)}$$

This is a most curious transformation. It is the quantum analogue of Sommerfeld's transformation, mentioned above in the classical relativistic Kepler problem.

In this frame

$$\Gamma \to \tilde{\Gamma} = S\Gamma S^{-1} = \beta K\left[1-\left(\frac{\alpha Z}{K}\right)^2\right]^{1/2} \tag{VII.14}$$

with the obvious eigenvalues

$$\tilde{\Gamma} \to \gamma\kappa = \pm\left[\kappa^2 - (\alpha Z)^2\right]^{1/2} \tag{VII.15}$$

Since the operator $i\frac{\vec{\alpha}}{2}$ is the generator of Lorentz boosts in the 4-dimensionsl space-time, the transformation S has the form of a finite boost, but with an operator valued angle. In the classical limit, one finds that the transformation S is a Lorentz transformation along an instantaneous direction tangent to the orbit having the velocity $\upsilon = c\frac{(\alpha Z)}{|K|}$. Such a Lorentz transformation would cause a time dilation between the fixed and transformed systems by a factor

$$\left\{1-\frac{\upsilon^2}{c^2}\right\}^{-1/2} = \left\{1-\left(\frac{\alpha Z}{K}\right)^2\right\}^{-1/2}$$

This is precisely the factor previously denoted by $1/\gamma'$ (see, Sec.II.3) in the classical limit.

In other words, one can interpret the transformation S to be the quantum mechanical analog to Sommerfeld's transformation to a system in which the relativistic orbits appear closed.

The use of this transformation brings the radial differential equation into non-relativistic form

$$\left(\frac{d^2}{dr^2}+\frac{2}{r}\frac{d}{dr}-\frac{l(\gamma)(l(\gamma)+1)}{r^2}-\frac{2\alpha ZE}{r}\right)u(r) = \left(E^2 - m^2\right)u(r) \tag{VII.16}$$

with the non-integer angular momentum $l(\gamma)$.

In the S frame the Johnson-Lippmann (JL) operator can be written as

$$A \to \tilde{A} = SAS^{-1} = \frac{1}{k}\beta\vec{\sigma}\hat{\vec{r}}\tilde{\Gamma}\left(i\hat{\vec{r}}\vec{p}+\frac{1+\tilde{\Gamma}}{r}+\frac{\alpha ZE}{\tilde{\Gamma}}\right) \tag{VII.17}$$

where $k=\left(E^2-m^2\right)^{1/2}$. This operator is not normalized to unity, because

$$\tilde{A}^2=\gamma^2+\frac{(\alpha ZE)^2}{k^2}$$

Thus a new normalization can be introduced

$$\tilde{A}'=\left(\gamma^2+\frac{(\alpha ZE)^2}{k^2}\right)^{-1/2}\tilde{A} \qquad \text{(VII.18)}$$

This operator factorizes into operators acting separately on the angular and radial components of the eigenkets. The radial part defines non-relativistic supercharges via

$$Q_\pm=\left(i\hat{\vec{r}}\vec{p}+\frac{1+\tilde{\Gamma}}{r}+\frac{\alpha ZE}{\tilde{\Gamma}}\right)\left[\frac{1}{2}(\rho_1\pm i\rho_2)\right] \qquad \text{(VII.19)}$$

where ρ_i are Dirac-Pauli matrices.

Then it is easy to check, that

$$\{Q_+,Q_-\}u(r)=\left(\frac{d^2}{dr^2}+\frac{2}{r}\frac{d}{dr}-\frac{l(\gamma)(l(\gamma)+1)}{r^2}-\frac{(\alpha ZE)^2}{\tilde{\Gamma}^2}-\frac{2\alpha ZE}{r}\right)u(r)=$$
$$=\left(E^2-m^2-\frac{(\alpha ZE)^2}{\gamma^2}\right)u(r) \qquad \text{(VII.20)}$$

Therefore we have a radial supersymmetry generated by JL operator in the Sommerfeld's frame. Relativity introduces two dynamical effects and correspondingly two extra interactions in the relativistic Coulomb problem with spin:

1) The quadratic Coulomb potential: $\left(\frac{\alpha Z}{r}\right)^2$,

2) A dynamical spin effect: $i(\vec{\alpha}\cdot\hat{\vec{r}})\frac{\alpha Z}{r^2}$

These two dynamical effects combine in such a way as to imply for the Dirac problem a precise analog to Sommerfeld's transformation to closed orbits. The net effect is to validate the structural symmetry of the non-relativistic calculation for the relativistic domain.

Let us remark that the Biedenharn approach has many other applications, for example, in Problem of the charged Dirac particle in a monopole field [7], dyons [8] and etc.

VII.2 Some Practical Generalizations: The LRL Vector in the Presence of an Electric Field [9]

When a unified electric field is present, the Schroedinger equation still separates in spherical and parabolic coordinates if the Z axis is in the direction of the electric field. So according of general view there must be some additional conserved quantity. Indeed:

Let us consider the equation of motion for a particle moving in an inverse square law force with an external electric field present

$$\frac{d}{dt}\vec{p} = -\frac{Ze^2}{r^2}\hat{\vec{r}} + e\vec{E} \tag{VII.21}$$

The angular momentum $\vec{L} = \vec{r} \times \vec{p}$ then satisfies the equation

$$\frac{d}{dt}\vec{L} = e\vec{r} \times \vec{E} \tag{VII.22}$$

The key to the following derivation is the observation that

$$\frac{d}{dt}\hat{\vec{r}} = \frac{1}{mr^2}\left[\vec{L} \times \hat{\vec{r}}\right] \tag{VII.23}$$

Now consider

$$\frac{d}{dt}\left[\vec{L} \times \vec{p}\right] = e\vec{L} \times \vec{E} - \frac{Ze^2}{r^2}\left[\vec{L} \times \hat{\vec{r}}\right] + e\left[\left[\vec{r} \times \vec{E}\right] \times \vec{p}\right] \tag{VII.24}$$

This may be compared to the time-derivative of the LRL vector, given by

$$\vec{A} = \hat{\vec{r}} - \frac{1}{Ze^2 m}\left[\vec{p} \times \vec{L}\right]$$

We have

$$\frac{d}{dt}\vec{A} = \frac{1}{Ze^2 m}\left\{e\left[\vec{L} \times \vec{E}\right] + e\left[\left[\vec{r} \times \vec{E}\right] \times \vec{p}\right]\right\} \tag{VII.25}$$

Therefore

$$\frac{d}{dt}\vec{A} \cdot \vec{E} = \frac{1}{Zem}\left[\vec{r} \times \vec{E}\right] \cdot \left[\vec{p} \times \vec{E}\right] = \frac{1}{2Ze}\frac{d}{dt}\left[\vec{r} \times \vec{E}\right]^2$$

So that

$$\frac{d}{dt}\left\{\vec{A}\cdot\vec{E}-\frac{1}{2Ze}\left[\vec{r}\times\vec{E}\right]^2\right\}=0$$

and the expression inside of parenthesis is a constant of the motion

$$\vec{A}\cdot\vec{E}-\frac{1}{2Ze}\left[\vec{r}\times\vec{E}\right]^2=const. \qquad \text{(VII.26)}$$

Consider the vector $\vec{C}$ given by

$$\vec{C}=\vec{A}-\frac{1}{2Ze}\left[\left[\vec{r}\times\vec{E}\right]\times\vec{r}\right] \qquad \text{(VII.27)}$$

then $\vec{C}\cdot\vec{E}$ is the above constant.

The vector $\vec{C}$ is a suitable generalization of $\vec{A}$ since it satisfies the simple equation of motion

$$\frac{d}{dt}\vec{C}=\frac{3}{2Zem}\left[\vec{L}\times\vec{E}\right]$$

The external field breaks the spherical symmetry, but there remains less (cylindrical) symmetry and we have a corresponding constant.

Above result and its derivation is valid quantum mechanically, provided care is taken with the order of the non-commuting operators as in the usual case.For this new constant the only change is that $\vec{C}$ must be symmetrized in the Weyl way

$$\vec{C}=\hat{\vec{r}}-\frac{1}{2Ze^2m}\left\{\left[\vec{p}\times\vec{L}\right]-\left[\vec{L}\times\vec{p}\right]\right\}-\frac{1}{2Ze}\left[\left[\vec{r}\times\vec{E}\right]\times\vec{r}\right] \qquad \text{(VII.28)}$$

This constant vector may have suitable application in the problems of Stark effect [10].

Conclusions

The main purpose of our book was to elucidate the symmetries related to the Kepler-Coulomb problem and describe the principal methods of their investigation. How this aim was attained is the privilege of our readers.

We have considered problems both in classical and quantum mechanics. The Kepler problem is one of the central problems in physics.By solving this problem in classical mechanics Newton demonstrated the fact that the motion of planets satisfies the three Kepler's laws.Needless to say, this is one of the most impessive and beautiful intellectual conquest of mankind.

After more then three centures, the Kepler problem still plays an important role in mathematics and physics. There has been a continuous interest in this problem. In particular, in the last three decades, we have witnessed an explosion of its interactions with quantum mechanics, celestial mechanics and mathematics. There were various generalizations of the Kepler problem in the last years. As we already know, according to the Bertrand theorem, the Kepler problem and harmonic oscillator are the only central dynamical systems which have closed orbits for all bounded motion.

The Kepler problem was generalizd to the MICZ-Kepler problem independently discovered by McIntosh and Cisneros [11] and Zwanziger[12] more than thirty years ago. Roughly speaking, MITZ-Kepler problem is the Kepler problem in the case when the nucleus of a hypothetic hydrogen atom also carries a magnetic charge (Dirac's monopole). In the phase space $\left(R^3 - \{0\}\right) \times R^3$ with Cartesian coordinates $\left(\vec{q}, \vec{p}\right)$, this problem has the Hamiltonian

$$H = \frac{\vec{p}^{\,2}}{2m} + \frac{\mu^2}{2mr^2} - \frac{a}{r}, \; \left(r = |\vec{q}|, \qquad a > 0, \qquad \mu = const.\right)$$

and admits the constants of motion consisting of the angular momentum and the LRL vector

$$\vec{J} = \vec{q} \times \vec{p} + \mu \frac{\vec{q}}{r}, \qquad\qquad \vec{A} = \vec{p} \times \vec{J} - ma\frac{\vec{q}}{r}$$

This implies that the orbits in the $\vec{q}$-space are conic sections, and therefore all the bounded orbits are closed. The symmetry property of the MICZ-Kepler problem has been studied in a series of articles [13-15].

The authors of this articles had shown that neither the harmonic oscillator nor the hydrogen atom retain their accidental degeneracy when the nucleus possesses a magnetic charge, but if a repulsive centrifugal potential proportional to the square of the pole strength is added, accidentally degenerate systems with a higher symmetry result. The charged Coulombic monopole has a $O(4)$ symmetry group generated by the total angular momentum together with a LRL vector constructed from the total angular momentum.

It is established [13] that there is an infinite number of Hamiltonian systems whose bounded motion are all periodic. However, these results do not contradict Bernard's theorem. The system Bernard treated has the Hamiltonian of the form

$$\frac{\vec{p}^{\,2}}{2m} + V(r),$$

in which $V(r)$ is a function to be determined so that the system may have the closed orbit property.

There are also models that generalize the MICZ-Kepler problem to the various directions: to the curved spaces[16], arbitrary potentials13] and etc. The generalized MITZ-Kepler system is defined by the following Hamiltonian [17]

$$H_{MICZ} = \frac{\vec{\pi}^2}{2m} + \frac{\mu^2}{2mr^2} + \frac{a}{r}$$

where $\pi_i = p_i + eA_i$ and

$$\left[\pi_i, \pi_j\right] = -\mu \frac{\varepsilon_{ijk} x_k}{r^3}, \qquad \left[\pi_i, x_j\right] = -\delta_{ij}$$

The distinguished peculiarity of this system is the close similarity with Coulomb problem, which lies in the existence of a hidden symmetry given by the angular momentum $\vec{J}$ and by the analog of LRL vector, which are defined by the expressions

$$\vec{J} = \vec{r} \times \vec{\pi} + \mu \hat{\vec{r}}, \qquad \vec{A} = \frac{1}{2m}\left[\vec{\pi} \times \vec{J} - \vec{J} \times \vec{\pi}\right] + a\hat{\vec{r}}$$

The hidden symmetry exists due to the appearance in the Hamiltonian of the specific centrifugal term $\mu^2 / 2mr^2$. The origin of this additional centrifugal term can be understand as follows [17]:

In the presence of monopole magnetic field the angular momentum of the system gets additional, spin-like term $\mu \hat{\vec{r}}$. The magnetic field of Dirac dyon is $\vec{M} = \frac{e}{2mc}\vec{J}$. Hence the interaction energy of this magnetic momentum with magnetic field is given by the expression

$$V_B = -\vec{M} \cdot \vec{B} = -\frac{e}{2mc}\vec{J} \cdot \frac{g\vec{r}}{r^3} = \frac{\mu^2}{2mr^2}$$

i.e. coincides with the centrifugal term in the MICZ-Kepler system.

It can be noted that MICZ-Kepler system also describes the relative motion of the Dirac dyons with electric and magnetic charges (e_1, g_1) and (e_2, g_2) with $e_1 g_2 - e_2 g_1 = c\mu$. Simiar to above mentioned case, the centrifugal term could be interpreted as the interaction energy of the induced dipole momentum with the dyon electric field plus interaction energy of the induced magnetic moment with the magnetic field.

It is evident that all these attempts need a generalization to the relativistic quantum mechanics. As regards to other attempts of the extension of these symmetry considerations, it is remarkable to note generalizations to arbitrary higher dimensions, both classically and quantum mechanically. Here first of all one has to extend the notion of the Coulomb potential. It is well-known that the theoretical concepts about the Coulomb potential are based on the Maxwell equations in classical theory and on the one-photon exchange mechanism in quantum electrodynamics.

In classical field theory it is famous Gauss' law, while in QED Coulomb potential follows after the retardation effects in the photon propagator are neglected. In both cases the Coulomb potential appears depending on space dimensionality, $1/r^{D-2}$. However the conserved quantity of kind of the LRL vector arises only when potential has the form $1/r$. Only in this case the bounded motion in classical mechanics is closed and periodic as was established by by P. Ehrenfest in 1920 [18]. By this reason he urgently insisted on 3-dimensionality of the physical space.

Concerning to higher dimensional spaces most of authors [19,20] just take the form $1/r$ as a Coulomb potential and prove the existence of conserved LRL vector.

Moreover in this case the hydrogen atom problem is solvable exactly by algebraic methods [21]. At the same time there appears N=2 superalgebra in quantum mechanics.

As regards to the Dirac equation there is conserved quantity analogous to the JL operator for $1/r$ potential in arbitrary dimensions [22].

The only serious problem in case of the Dirac equation is the extension of γ^5 matrix to odd and even dimensional spaces. As was shown in [22] one can define γ^5 formally as a matrix γ^{D+1} require only anticommutativity with all other Dirac's γ-matrices.

Therefore as follows from our considerations above by requirement of N=2 supersymmetry one can prove that the arbitrary central potential $V(r)$ would have $1/r$ behaviour. So only for such potential have we an "accidental" supersymmetry.

REFERENCES

[1] Biedenharn L.C. Remarks on the Relativistic Kepler Problem. *Phys.Rev.* 126, 845 (1961).

[2] Biedenharn L.C., Swany N.V., Remarks on the Relativistic Kepler Problem.II. Approximate Dirac Coulomb Hamiltonian Possessing two Vector Invariants. *Phys. Rev.* B133, 1225 (1969).

[3] Biedenharn L.C. The Sommerfeld Puzzle, *Fund. of Physics*, 13, 13 (1983).

[4] Martin P.C. and Glauber R.F.Relativistic Theory of Radiative Orbital Electron Capture. *Phys.Rev.*109, 1307 (1958).

[5] Dahl J.P. and Jorgensen T. On the Dirac-Kepler Problem: The Johnson-Lippmann Operator, Supersymmetry, and Normal-Mode Representations. *Int. J. Quant. Chem.* 53, 161 (1995).

[6] Temple G. *The General Principles of Quantum Mechanics.* New York, 1948.

[7] Horvathy P.A. *The Biedenharn Approach to Relativistic Coulomb Type Problems.* arXiv: hep-th/0601123 (2006) and References therein.

[8] Stahlhofen A.A. On the Dyon Problem. *Mod. Phys. Lett.* A5, 2007 (1990).

[9] Redmond P.J. Generalization of the Runge-Lenz Vector in the Presence of an ElectricField. *Phys.Rev.*B133,1352(1964).

[10] Valent G. The hydrogen Atom in Electric and Magnetic fields: Pauli's 1926 Article. *Am.J.Phys.*71, 171 (2003)

[11] McIntosh H.V. and Cisneros A. Degeneracy in the Presence of a magnetic Monopole. J.Math.Phys 11,896(1970).

[12] Zwanziger D. Exactly Soluble Nonrelativistic Model of Particle with both Electric and Magnetic Charges. *Phys.Rev.* 176, 1480 (1968).

[13] Iwai T. nd Katayama N. Multifold Kepler Systems-Dynamical Systems all of Whose Bounded Trajectories are Closed. *J.Math.Phys.*, 36, 1790 (1995).

[14] Meng G. *The MICZ-Kepler Problem in All Dimensions.* arXiv: math-ph/0507028 (2005).

[15] Meng G. *Generalized Kepler Problem.* aXiv: math-ph/0509002 (2005)

[16] Cataescu I. and Visinescu M. *Hierarxhy of Dirac, Pauli and Klein-Gordon Conserved Operators in Taub-NUT background.* arXiv: hep-th/0107205 (2001).

[17] Nersessian A. *Generalization of MICZ-Kepler System.* arXiv: 0711.1037 [math-ph], Nov. 2007

[18] Ehrenfest P. *Ann. Phys.* 61, 440 (1920).

[19] Gu X.-Yan et al. Exact Solutions to the Dirac Equation for a Coulomb Potential in D+1 Dimensions. *It. J. Mod. Phys.* E11, 335 (2002).

[20] Wipf A. et al. Algebraic Solution of the Supersymmetric Hydrogen Atom. *Quantum Theory and Symmetries* IV. Sofia, 2006.

[21] Kirchberg et al. Algebraic Solution of the Supersymmetric Hydrogen Atom in d imensions. *Annals Phys.* 303, 359 (2003).

[22] Katsura H. and Aoki H. Exact Supersymmetry in the Relatvistic Hydrogen Atom in General Dimensions – Supercharge and the Generalized Johnson-Lippmann Operator. Math. Phys. 47, 032302 (2006).

BIBLIOGRAPHY (PART I)

This part of Bibliography is borrowed from the excellent Review article by Harold V. McIntosh "SYMMETRY AND DEGENERACY", which contains rather complete citations up until this paper was edited (1970). We only arrange it in alphabetical order. List of cosequent articles is done after this part below.

Alexandrow W., Die magnetishe Ablenkung der Korpuskularstrahlen in der Diracschen Wellenmechanik. *Ann. Physik* (5) 2, 477-484 (1929).

Alliluev S. P. and Matveenko A. V., Symmetry group of the hydrogen molecular ion. Soviet Physics *JETP* 24, 1260-1264 (1967).

Alliluev S. P., On the relation between "accidental" degeneracy and "hidden" symmetry of a system. Soviet Phys. *JETP* 6, 156-159 (1959).

Bacry H. and Ruegg H., Dynamical groups and spherical potentials in classical mechanics. *Comm. Math. Phys.* 3, 323-333 (1966).

Baker G. A. Jr., Degeneracy of the n-dimensional isotropic, harmonic oscillator. *Phys. Rev.* 103, 1119-1120 (1956).

Banderet P. P., Zur Theorie der singularen Magnetpole. *Helv. Phys. Acta* 19, 503-522 (1946).

Bargmann V., Zur Theorie des Wasserstoffatoms. *Z. Physik* 99, 576-582 (1936).

Berrondo M. and McIntosh H. V., Degeneracy of the Dirac equation with electric and magnetic Coulomb potentials. *J. Math. Phys.* 11, 125-141 (1970).

Bertrand J., Theoreme relative au mouvement d'un point attire vers un centre fixe. C. R. *Acad. Sci. Paris* 77, 849-853 (1873).

Bethe H. A. and Maximon L. C., Theory of Bremsstrahlung and pair production I. differential cross section. *Phys. Rev.* 93, 768-784 (1954).

Bethe H. A. and Salpeter E. E., "Quantum Mechanics of One- and Two-Electron Atoms." Academic Press, New York, 1957.

Biedenharn L. C., Remarks on the relativistic Kepler problem. Phys. Rev. 126, 845-851 (1962). MA11, symmetry considerations in the Dirac-Coulomb problem. *Bull. Amer. Phys. Soc.* 7, 314(A)(1962).

Biedenharn L.C. and Swamy N.V.V.J. Remarks on the RelativisticKepler ProblemII, Approximate Dirac Hamiltonian Having Two Vector Invariants. *Pys.Rev.* B126, 845-851 (1964).

Born M., "The Mechanics of the Atom." Ungar, New York, 1960.

Breit G., An interpretation of Dirac's theory of the electron. Proc. Nat. Acad. Sci. U. S. A. 14, 553-559 (1928); On the interpretation of Dirac's α -matrices. *Proc. Nat. Acad. Sci. U. S. A.* 17, 70-73 (1931).

Brown L. M., Two component fermion theory. Phis. Rev. 111, 957-964 (1958).

Burkhard D. G., Factorization and wave functions for the symmetric rigid rotator. *J. Mol. Spectry.* 2, 187-202 (1958).

Case K. M., Singular potentials. *Phys. Rev.* 80, 797-806 (1950).

Casimir H. G. B., Zur quantenmechanischen Behandlung des Kreiselproblems. *Z. Physik* 59, 623-634(1930).

Cini M. and Touscheck B., The relativistic limit of the theory of spin $\frac{1}{2}$ particles. Nuovo Cimento 7, 422-423 (1958).

Cisneros A. and McIntosh H. V., Symmetry of the two-dimensional hydrogen atom. *J. Math. Phys.* 10, 277-286 (1969).

Cisneros A., Estudio sobre el gruo universal de simetria. Professional Thesis, Polytech. Inst. Of Mexico, 1968; A. Cisneros and H. V. McIntosh, Search for a universal symmetry group in two dimensions. *J. Math. Phys.* 11, 870-895 (1970).

Coulson C. A. and Joseph A., A constant of the motion of the two-center Kepler problem. *Intern. J. Quantum Chem.* 1, 337-347 (1967).

Coulson C. A. and Joseph A., Self-adjoint ladder operators II. *Rev. Modern Phys.* 39, 838-849 (1967).

Darboux G., Etude d'une question relative au mouvement d'un point sur une surface de revolution. *Bull. Soc. Math. France* 5, 100-113 (1877).

Darboux G., Recherche de la loi que doit suiver une force centrale pour que la trajectoire qu'elle determine soit toujours une conique. *C. R. Acad. Sci. Paris* 84, 936-938 (1877).

Darwin C. G., On the diffraction of the magnetic electron. *Proc. Roy. Soc. Ser.* A 120, 631-642 (1928).

Demkov Yu. N., Group symmetry of an isotropic oscillator. Zh. Eksperim. Teor. Fiz. 26, 757 (1954); Symmetry group of the isotropic oscillator. Soviet Phis. JETP 9, 63-66 (1959); Bull. Leningrad State Univ. 11, 127 (1953); The definition of the symmetry group of a quantum system. The anisotropic oscillator. Soviet Phys. *JETP* 17, 1349-1351 (1963).

Dennison D. M., The rotation of molecules. *Phys. Rev.* 28, 318-333 (1926).

Dirac P. A. M., Quantized singularities in the electromagnetic field. *Proc. Roy. Soc. Ser.* A 133, 60-72 (1931).

Dulock V. A. and McIntosh H. V., Degeneracy of cyclotron motion. *J. Math. Phys.* 7, 1401-1412 (1966).

Dulock V. A. and McIntosh H.V., On the degeneracy of the Kepler problem. *Pacific J. Math.* 19, 39-55 (1966).

Dulock V. A., Jr., A study of accidental degeneracy in Hamiltonian mechanics. Ph. D. Thesis, Univ. of Florida, 1964.

Eldridge J. A., Strings, poles, and the electron. *Phys. Rev.* 75, 1614-1615 (L) (1949).

Eliezer C. J. and Roy S. K., The effect of a magnetic pole on the energy levels of a hydrogen-like atom. *Proc. Cambridge Philos. Soc.* 58, 401-404 (1962).

Elliott J. P., Collective motion in the nuclear shell model I. Classification schemes for states of mixed configurations. *Proc. Roy. Soc. Ser.* A 245, 128-145 (1958).

El-Sherbini M. A., Three-dimensional periodic orbits in the field of a non –neutral atom. *Philos. Mag.* 14, 304-310 (1932).

Epstein E., Bermerkungen zur Frage der Quantelung des Kreisels. *Physik. Z.* 20, 289-294 (1919).

Erikson H. A. and Hill E. L., A note on the one-electron states of diatomic molecules. *Phys. Rev.* 75, 29-31 (1949).

Feynman R. P. and Gell-Mann M., Theory of the Fermi interaction. *Phys. Rev.* 109, 193-198 (1958).

Fierz M., Zur Theorie magnetisch gladdener Teilchen. *Helv. Phys. Acta* 17, 27-34 (1944).

Fock V. A., Zur Theorie des Wasserstoffatoms. *Z. Physik* 98, 145-154 (1935); see also V. A. Fock, Wasserstoffatom und nicht-euklidishe Geometrie (mit einer Deutshen Zusammenfassung). *Izv. Akad. Nauk USS*R 2, 169-188 (1935).

Foldy L. and Wouthuysen S. A., On the Dirac spin $\frac{1}{2}$ particle and its nonrelativistic limit. *Phys. Rev.* 78, 29-36 (1950).

Ford K. W. and Wheeler J. A., An application of semiclassical scattering anlysis. *Ann. Physics* 7, 287-322 (1959).

Fradkin D. M., Existence of the dynamical symmetries O(4) and SU(3) for all classical central potential problems. *Progr. Theoret. Phys.* 37, 798-812(1967).

Galindo A. and C. Sanchez del Rio, Intrinsic magnetic moment as a non-relativistic phenomenon. *Amer. J. Phys.* 29, 582-584 (1961).

Gibbs J. W. and Wilson E. B., "Vector Analysis." Scribners and Sons, New York, 1901 Reprinted by Yale Univ. Press, New Haven, Connecticut, 1958.

Goldhaber A. S., Role of spin in the monopole problem. *Phys. Rev.* 140, 1407-1414 (1965).

Goshen S. (Goldstein) and Lipkin H. J., A simple independent particle system having collective properties. Ann. Physics 6, 310-309 (1959); A simple model of a system possessing rotational states. *Ann. Physics* 6, 310-318 (1959).

Goto E., On the observation of magnetic poles. J. Phys. Soc. Japan 13, 1413-1418 (1958); Expected behaviour of the Dirac monopoe in cosmic space. Prog. Theoret. Phys. 30, 700-718 (1963); E. Goto, H. H. Kolm, and K. W. Ford, Search for ferromagneticalliy trapped magnetic monopoles of cosmic origin. *Phys. Rev.* 132, 387-396 (1963).

Greenberg D. F., Accidental degeneracy. *Amer. J. Phys.* 34, 1101-1109 (1966).

Greenhill G. (Sir), Orbits in the Field of a doublet and generally of two centres of force. *Philos. Mag.* 46, 364-385 (1923).

Gronblom B. O., Uber singulare Magnetpole. Z. Physik 98, 283-285 (1935).

Hamilton W., On the application of the method of quaternions to some dynamical questions. Proc. Roy. Irish Acad. 3, 441-448(1847); see also (Sir) W. R. Hamilton, "Elements of Quaternions." Longmans, Green, New York, 1866.

Han M. Y. and Stehle P., SU(2) as a classical invariance group. Nuovo Cimento 48, 180-187 (1967).

Harish-Chandra, Motion of an electron in the field of a magnetic pole. *Phys. Rev.* 74, 883-887 (1948).

Higab M. A., Three-dimensional motion of an electron in the field of a non-neutral atom. Philos. Mag. 7, 31-52 (1929). Two-dimensional periodic orbits in the field of a non-neutral[atom]. Philos.Mag. 7, 783-792 (1929). Periodic orbits in a field of force defined by a certain potential. *Philos. Mag.* 14, 298-304 (1932).

Hill E. L., Seminar on the theory of quantum mechanics. Univ. of Minnesota, 1954 (unpublished).

Huff L. D., The motion of a Dirac electron in a magnetic field. *Phis. Rev.* 38, 501-512 (1931).

Hulthen L., Uber die quantenmechanische Herleitung der Balmerterme. *Z. Physik* 86, 21-23 (1933).

Hund F., Bemerkung uber die Eigenfunktionen des Kugelkreisels in der quantenmechanic. *Z. Physik* 51, 1-5 (1928).

Hwa R. C. and Nuyts J., Group embedding for the harmonic oscillator. *Phys. Rev.* 145, 1188-1195 (1966).

Hylleraas E. A. Die Wellengleichung des Keplerproblems im Impulsraum. *Z. Physik* 74, 216-224 (1932).

Hylleraas E. A., Zur practischen Losung der relativistischen Einelektronengleichungen. Z. Phisik 140, 626-631 (1955); Zur practischen Losung der relativistischen Einelektronengleichungen.II. *Z. Phisik* 164, 493-506 (1961).

Il'kaeva L. A., Symmetry group of the anisotropic oscillator. Vestnik Leningrad. Univ. 22, 56-62 (1963)(in Russian).

Infeld L. and Hull T. E., The factorization method. Rev. *Modern Phys.* 23, 21-68 (1951).

Ionesco-Pallas N. J., Relativistic wave equation for plane motion. *Rev. Roumaine Phys.* 12, 327-336 (1967).

Jauch J. M. and Hill E. L., On the problem of degeneracy in quantum mechanics. P*hys. Rev.* 57, 641-645 (1940).

Jauch J. M., Groups of quantum mechanical contact transformations and the degeneracy of energy levels. *Phys. Rev.* 55, 1132 (A)(1939).

Jauch J. M., On contact transformations and group theory in quantum mechanical problems. Ph. D. Thesis, Univ. of Minnesota, 1939.

Jeans J. H., On the motion of a particle about a doublet. *Philos. Mag.* 20, 380-382 (1910).

Johnson M. H. and Lippmann B. A., Motion in a constant magnetic field. *Phys. Rev.* 76, 828-832 (1949).

Johnson M. H. and Lippmann B. A., Relativistic Kepler problem. *Phys. Rev.* 78, 329 (A) (1950).

Johnson M. H. and Lippmann B. A., Relativistic motion in a magnetic field. *Phys. Rev.* 77, 702-705 (1950).

Jordan P., Uber die Diracschen Magnetpole. *Ann. Physik* 32, 66-70 (1938).

Kennard E. H., Zur Quantenmechanik einfacher Bewegungstypen. *Z. Physik* 44, 326-352 (1927).

King G. W., Hainer R. M., and Cross P. C., The asymmetric rotor I: Calculation and symmetry classification of energy levels. *J. Chem. Phys.* 11, 27-42 (1943)

Klein F. and Sommerfeld A., "Theorie des Kreisels." Johnson Reprint Corp., New York, 1965.

Klein O. Die reflexion von Electrons im homogenen elektrischen Feld nach der relativistischen Dynamik von Dirac. *Z. Physik* 53, 157-165 (1929).

Klein O., Zur Frage der Quantelung des symmetrishen kreisels. *Z. Physik* 58, 730-734 (1929).

Kolossoff G., Uber Behandlung zyklischer systeme mit Variationsprinzipien mit Anwendungen auf die Mechanik starrer Korper. *Math. Ann.* 60, 232-240 (1905).

Kolsrud M., On the solution of Dirac's equation with Coulomb potential. *Phys. Norveg.* 2, 43-50 (1966).

Krammers H. A. and Ittman G. P., Zur quantelung des asymmetryshen Kreisels. Z. Physik 53, 553-565 (1929); Zur quantelung der asymmetryshen Kreisels,II. Z. Physik 58, 217-231 (1929); Zur quantelung der asymmetryshen Kreisels,III. *Z. Physik* 60, 663-681 (1930).

Krammers H. A. and Pauli W. Jr., Zur Theorie der Bandspektern. *Z. Physik* 13, 351-367 (1923).

Krammers H. A., Uber die Quantelung rotierender Molekule. *Z. Physik.* 13, 343-350 (1923).

Kronig R. de L. and Rabi I. I., The symmetrical top in the undulatory mechanics. *Phys. Rev.* 29, 262-269 (1927).

Landau L., diamagnetismus der Metalle. *Z. Physik* 64, 629-637 (1930).

Lapidus J. R. and Pietenpol J. L., Classical interaction of an electric charge with a magnetic monopole. *Amer. J. Phys.* 28, 17-18 (1960).

Laporte O. and Rainich G. Y., Stereographic parameters and pesudominimal hypersurfaces. *Trans. Amer. Math. Soc.* 39, 154-182 (1936).

Laporte O., The approximation of geometric optics as applied to a Dirac electron moving in a magnetic field. *Phys. Rev.* 42, 340-347 (1932).

Laporte O., The connection between the Kepler problem and the four dimensional rotation group. *Phys. Rev.* 50, 400(A)(1936).

Lehnert B., "Dynamics of Charged Particles." North-Holland Publ., Amsterdam, 1964.

Lehti R., Some special types of perturbative forces acting on a particle moving in a central field. *Ann. Acad. Sci. Fenn. Ser.* A 6, No. 282 (1968).

Lenz W., Uber den Bevegungsverlauf und die Quantenzustande der gestorten Keplerbevegung. *Z. Physik* 24, 197-207 (1924).

Levy-Leblond J. M., Galilean quantum field theories and a ghostless Lee model. Comm. Math. Phys. 4, 157-176 (1967); Nonrelativistic particles and wave equations. Comm. *Math. Phys.* 6, 286-311 (1967).

Loudon R., One- dimensional hydrogen atom. *Amer. J. Phys.* 27, 649-655 (1959).

Luming M. and Predazzi E., Theory of O(4) invariant potentials and its application to a short-range example. Ann. Physics 40, 221-236 (1966); A set of exactly soluble potentials. *Nuovo Cimento* B 41, 210-212 (L) (1966).

Lutgemeier F., Zur Quantentheorie des drei- und mehratomigen Molekuls. *Z. Physik* 38, 251-263 (1926).

Maiella G. and Vilasi G., Redusible representations of the symmetry group of the anisotropic harmonic oscillator. Lett. *Nuovo Cimento* 1, 57-64 (1969).

Maiella G. and Vitali B., A note on dynamical symmetries of classical systems. *Nuovo Cimento* A 47, 330-333 (1967).

Malkus W. V. R., The interaction of the Dirac monopole with matter. *Phys. Rev.* 83, 899-905 (1951).

Martin P. C. and Glauber R.J., Relativistic theory of radiative orbital electron capture. *Phys. Rev.* 109, 1307-1325 (1958).

McIntosh H. V., Degeneracy of the magnetic monopole. Bull. Amer. Phis. Soc. 12, 699(A) (1967); H. V. McIntosh and A. Cisneros, Motion of a charged particle in the field of a magnetic monopole. *Bull. Amer. Phys. Soc.* 13, 909 (A) (1968);

McIntosh H. V., On accidental degeneracy in classical and quantum mechanics. *Amer. J. Phys.* 27, 620 625 (1959).

Miller W. Jr., On Lie algebras and some special functions of mathematical physics. Mem. Amer. Math. Soc. No. 50 (1964); "Lie Theory and Special Functions." Academic Press,

New York, 1968; Weisner L., Group theoretic origin of certain generating functions. Pacific J. Math 5, 1033-1039 (1955); H. R. Coish, Infeld factorization and angular momentum. Canad. *J. Phys.* 34, 343-349 (1956).

Moshinsky M., Soluble many-body problem for particles in a Coulomb potential. *Phys. Rev.* 126, 1880-1881 (1962).

Mukunda N., Realizations of Lie algebras in classical mechanics. J. Math. Phys. 8, 1069-1072 (1967); Dynamical symmetries and classical mechanics. Phys. Rev. 155, 1383-1386 (1967); N. Mukunda, L. O'Raifeartaigh, and E. C. G. Sudarshan, Characteristic noninvariance groups of dynamical systems. *Phys. Rev. Lett.* 15, 1041-1044 (1965).

Nadeau G., Concerning the classical interaction of an electric charge with a magnetic monopole. *Amer. J. Phys.* 28, 566 (L) (1960).

Newton T. D. and Wigner E. P., Localized states for elementary systems. *Rev. Modern phys.* 21, 400-406 (1949).

Nikolsky K., Das Oszillatorproblem nach der Diracshen Theorie. *Z. Physik* 62, 677-681(1930).

O'Connell R. F., Motion of a relativistic electron with an anomalous magnetic moment in a constant magnetic field. *Phys. Lett.* A 27, 391-392 (1968).

Page L., Deflection of electrons by a magnetic field in the wave mechanics. *Phys. Rev.* 36, 444-456 (1930).

Page L., Three-dimensional periodic orbits in the field of a non –neutral dipole. *Philos. Mag.* 10, 314-323 (1930).

Pauli W. Jr., Uber das Model des Wasserstoffmolekulions. *Ann. Physik* 68, 177-240 (1922).

Pauli W. Jr., Uber das Wasserstoffspektrum vom Standpunkt der neuen Quanten-mechanick. *Z. Physik* 36, 336-363 (1926).

Pauling L. and Wilson E. B. Jr., "introduction to quantum Mechanics," §14. McGraw-Hill, New York, 1935.

Peres A., Singular strings of magnetic monopoles. *Phys. Rev. Lett.* 18, 50-51 (1967); Rotational invariance of magnetic monopoles. *Phys. Rev.* 167, 1449 (1968).

Plesset M. S., Relativistic wave mechanics of electrons deflected by a magnetic field. *Phys. Rev.* 36, 1728-1731 (1930).

Podolsky D. and Pauling L., The momentum distribution in hydrogenlike atoms. *Phis. Rev.* 34, 109-116 (1929).

Poincare H., Remarques sur une experience de M. Birkeland. C. R. *Acad. Sci.* Paris 123, 530-533 (1893).

Postepska I., Harmonischer Oszillator nach der Diracschen Wellengleichung. *Acta Phys. Polon.* 4, 269-280 (1935).

Pryce M. H. L., The mass-centre in the restricted theory of relativity and its connexion with the quantum theory of elementary particles. *Proc. Roy. Soc. Ser.* A 195, 62-81(1949).

Rabi I. I., Das freie Elektron im homogenen Magnetfield nach der Diracshen Theorie. *Z. Physik* 49, 507-511 (1928).

Ramsey N. F., Time reversal, charge conjugation, magnetic pole conjugation, and parity. *Phys. Rev.* 109, 225-226 (L) (1958).

Ravenhall D. G., Sharp R. T. and Parde W. J., A connection between SU(3) and O(4). *Phys. Rev.* 164, 1950-1956 (1967).

Ray B. S., Uber die Eigenwerte des asymmetryschen Kreisels, *Z. Physik* 78, 74-91 (1932).

Redmond P. J., Generalization of the Runge-Lenz vector in the presence of an electric field. *Phys. Rev.* B 133, 1352-1353 (1964).

Reiche F., Zur Quantelung des asimetrishen Kreisels. *Physik. Z.* 19, 394-399 (1918).

Riche F. and Hans Rademacher, Die Quantelung des symmetrishen Kreisels nach Schrodingers Undulationsmechanik. *Z. Physik* 39, 444-464 (1926).

Rosen J., On realizations of Lie algebras and symmetries in classical and quantum mechanics.. *Nuovo Cimento* A 49, 614-621 (1967).

Runge C., "Vectoranalysis," English transl. Dutton, New York, 1919.

Saenz A. W., On integrals of motion of the Runge type in classical and quantum mechanics. Ph. D. Thesis, Univ. of Michigan, 1949.

Saha M. N., Note on Dirac's theory of magnetic poles. *Phys. Rev.* 75, 1968 (L) (1949).

Saha M. N., The origin of mass in neutrons and protons. *Indian J. Math.* 10, 141-153 (1936).

Sauter F. Uber das Verhalten eines Electrons im homogenen elektrischen Feld nach der relativistischen Theorie Dirac's. *Z. Physik* 69, 742-764 (1931).

Sauter F., Losung der Diracshen Gleichungen ohne Spezialisierung der Diracschen Operatoren. Z. Physik 63, 803-814 (1930); Zur Losung der Diracshen Gleichungen ohne Spezialisierung der Diracschen Operatoren, II. Z. Physik 64, 295-303 (1930);); Zur Losung der Diracshen Gleichungen fur ein zentralsymmetrisches Kraftfeld. *Z. Physik* 97, 777-784 (1935).

Schrodinger E., A method of determining quantum-mechanical eigenvalues and eigenfunctions. Proc. Roy. Irish Acad. Sect. A 46, 9-16 (1940); Further studies on solving eigenvalue problems by factorization. *Proc. Roy. Acad. Sect.* A 46, 183-206 (1941)

Schrodinger E., Uber die kraftefreie Bewegung in der relativistishen Quantenmechanik. Berliner Ber. pp. 418-428 (1930); Zur Quantendynamik des Elektrons. Berliner Ber. pp. 63-72 (1931).

Schwinger J. "On Angular Momentum" (L. C. Beidenharn and H. Van Dam, eds.), pp. 229-279. Academic Press, New York, 1965.

Schwinger J., Magnetic charge and quantum field theory. Phys. Rev. 144, 1087-1093 (1966); Sources and magnetic charge. *Phys. Rev.* 173, 1536-1554 (1968).

Sexl T., Bemerkung zur Quantelung des harmonishen Oszillator im Magnetfelde. Z. Physik. 48, 611-613 (1928); Fock V., Bemerkung zur Quantelung des harmonishen Oszillators im Magnetfeld. *Z. Physik* 47, 446-448 (1928).

Shaffer W. H., Operational derivation of wave functions for a symmetrical rigid rotator. *J. Mol. Spectry.* 1, 69-80 (1957);

Sheth C. V. The relativistic symmetric Hamiltonian and the Foldy-Wouthuysen transformation. *Nuovo Cimento* A 54, 549-551 (1968).

Sommerfeld A. and Maue A. W., Approximate adaptation of a solution of the Schrodinger equation to the solution of Dirac's equation. *Ann. Physik* 22, 629- 642 (1935).

Sommerfeld A. and Welker H., Kunstliche Grensbedigungen beim Keplerproblem. *Ann. Physik* 32, 56-65(1938).

Sommerfeld A., Zur Quantentheorie der Spectrallinien. *Ann. Physik* 51, 1-94, 125-167 (1916).

Stanciu G. N., Further exact solutions of the Dirac equation. *J. Math. Phys.* 8, 2043-2047 (1967).

Stehle P. and Han M. Y., Symmetry and degeneracy in classical mechanics. *Phys. Rev.* 159, 1076-1082 (1967).

Stevenson F., Note on the "Kepler problem" in a spherical space and the factorization method of solving eigenvalue problems. *Phys. Rev.* 59, 842-843 (L) (1941);

Stormer C., Periodische Elektronenbahnen im Felde eines Elementarmagneten und ihre Anwendung auf Bruches Modellversuche und auf Eschenhagens Elementarwellen des Erdmagnetismus. Z. Astrophys. 1, 237-274 (1930); Ein Fundamentalproblem der Bewegung einer elektrisch geladenen Korpuskel im kosmischen Raume. Z. Astrophys. 3, 31-52 (1931); Ein Fundamentalproblem der Bewegung einer elektrisch geladenen Korpuskel im kosmischen Raume II. Z. Astrophys. 3, 227-252 (1931); Ein Fundamentalproblem der Bewegung einer elektrisch geladenen Korpuskel im kosmischen Raume III. Z. Astrophys. 4, 290-318 (1932); "The Polar Aurora." Oxford Univ. Press (Clarendon), London and New York, 1955.

Susskind L. and Glogower J. Quantum mechanical phase and time operator. Physics 1, 49-61 (1964); P. Carruthers and M. M. Nieto, Coherent states and the forced quantum oscillator. *Amer. J. Phys.* 33, 537-544 (1965).

Tamm I., Die verallgemeinerten Kugelfunktionen und die Wellenfunktionen eines Elektrons im Felde eines Magnetpoles. Z. Physik 71, 141-150 (1931).

Temple G., The operational wave equation and the energy levels of the hydrogen atom. *Proc. Roy. Soc. Ser.* A 127, 349-360 (1930). "The General Principles of quantum Theory." Methuen, London, 1956.

Thomson J. J. (Sir), On the theory of radiation. Philos. Mag. 20, 238-247 (1910).

Uhlenbeck G. E. and Young L. A., The value of e/m by deflection experiments. *Phys. Rev.* 36, 1721-1727 (1930).

Van der Waerden B. L., "Sources of Quantum mechanics," p. 58. North-Holland Publ., Amsterdam, 1967.

Vendramin I. On the dynamical symmetry of the nonisotropic oscillators. *Nuovo Cimento* A 54, 190-192 (1968).

Wang S. C., On the asymmetric top in quantum mechanics. *Phys. Rev.* 34, 243-252 (1929)

Wentzl G. Comments on Dirac's theory of magnetic monopoles. *Theoret. Phys. Suppl.* 37 and 38, 163-174 (1966).

Whittaker E. T., "A treatise on the Analytical Dynamics of Particles and Rigid Bodies," 4th ed. Cambridge Univ. Press, London and New York, reprinted 1961.

Wigner E. and J. von Neumann, Behaviour of eigenvalues in adiabatic processes. *Physik. Z.* 30, 467-470 (1929).

Wigner E. and J. von Neumann, Uber merkwurdige diskrete Eigenwerte. *Physik.* Z. 30, 465-467(1929).

Wilson H. A., Note on Dirac's theory of magnetic poles. Phys. Rev. 75, 309(L) (1949).

Witmer E. E., the quantization of the rotational motion of the polyatomic molecule by the new wave mechanics. *Proc. Nat. Acad. Sci. U.S.A.* 13, 60-65 (1927).

Zwanziger D., Exactly soluble nonrelativistic model of particles with both electric and magnetic charges. Phys. Rev. 176, 1480-1488 (1968); Quantum field theory of particles with both electric and magnetic charges. *Phys. Rev.* 176, 1489-1495 (1968).

PART II

Alavi S.A. *Lamb Shift and Stark Effect in simultaneous space-space and momentum-momentum noncommutative quantum mechanics and theta-deformed su(2) algebra.* arXiv:hep-th/0501215.

Alhaidari A.D. Graded Extension of so(2,1) Lie Algebra and the Search for Exat Solutions of the Dirac Equation by Point Canonical Transfrmations. *Phys. Rev.*A65, 042109 (2004).

Argueso F. and Sanz J.L. Post-Newtonian Extensions of the Runge-Lenz Vector. *J.Math.Phys.* 25, 293 (1984).

Atom. arXiv: math-ph/0408012.

Auvil P.R. and Brown L.M. *The Relativistic Hydrogen Atom: A Simple Solution. Am.J.Phys.* 46, 679 (1978).

Backry H. and Richard J.L. Partial Group-Theoretical Treatment for the Relativistic Hydrogen Atom. *J. Math.Phys.* 8, 223) (1967).

Bacry H et al. Dynamical Groups and Spherical Potentials in Classical Mechanics. Comm. *Math. Phys.* 3. 323 (1996).

Bacry H. Comment on "The Relationship Between the Symmetries of and the existence of Conserved Vectors for the Equation $\ddot{\vec{r}} + f(r)\vec{L} + g(r)\vec{r} = 0$. *J.Phys. A: Math. Gen.* 24, 1157 (1991).

Bacry H. et al. Dynamical Groups and Spherical Potentials in Classical Mechanics. Commun. *Math. Phys.* 3, 323 (1966).

Baeyer H.C. Semiclassical quantization of the Relativistic Kepler Problem. *Phys.Rev.*d12, 3086 (1971).

Bartnik et al. Equation of Motion Using the Dynamical Evolution of he Runge-Lenz Vector. *Astrophys. Journ.* 34, 517 (1988).

Bates L. *Symmetry Preserving Deformations of the Kepler Prblem,* Preprint, Univ. Calgary, 1990.

Bawin M. SO(2.1) and Strong Coulomb Coupling. *J.Phys.A: Math.Gen.* 10, 1 (1977).

Benitez J. et al. Solution and Hidden Supersymmetry of a Dirac Oscillator. *Phys. Rev.* Lett. 64, 1643 (1990).

Bhattacharya K. *Solution of the Dirac Equation of an uniform Magnetic Field.* arXiv:0705.4275[hep-th].

Biedenharn L.C. and Louck L.D. *Angular Momentum in Quantum Physics. In "Encyclopedia of Mathematics and Its Application".* Addison-Wisley. 1981.

Biedenharn L.C. The "Sommerfeld Puzzle" Revisited and Resolved. *Found. Phys.* 13, 13 (1983).

Boyer T.H. Unfamiliar tTrajectories for a Relativitic Particle in a Kepleror Coulomb Potential. *Am.J.Phys.* 72, 992 (2004).

Brihaye Y. and Nininahazwe A. *Dirac Oscillators and Quasi-Exactly Solvable Operators.* arXiv: quant-ph/0503137.

Bruce S. and Roa L. A Dirac Particle in a Vector-like Hydrogenic Potential. *Nuovo Cim.*111, 159 (1998).

Buch L.H. and Denman H.H. Conserved and Piecewise Conserved Runge Vectors for the Isotropic Harmonic Oscillator. *Am.J.Phys.* 43,1046 (1975).

Burkhardt C.E. and Leventhal J.J. Lenz Vector Operations n Spherical Hydrogen Atom Eigenfunctions. *Am.J.Phys.* 72, 1013 (2004).

Carter B. and McLenaghan R.G. Geeralized Total Angular Momentum Operator for the Dirac Equation in Crved Space-Time. *Phys.Rev.* D19, 1093 (1979).

Carter B. Killing Tensor Quantum Numbers and Conserved Currents in Curved Space. *Phys. Rev.* D16, 3395 (1977).

Caspar Max. *Kepler.*Dover, New York.1993.

Chaffin E.F. Group Structure of the Relativistic Equivalent Oscillator. *J.Math. Phys.* 14, 977 (1973).

Chaichian M. et al. *Hydrogen Atom Spectrum and the Lamb Shift in Noncommutative QED.* arXiv:hep-th/0010175.

Cllinson C.D. Invetigation of Planetary Orbits Using the Lenz-Runge Vector. *Bull.Inst.Math.Appl.* 9,377 (1973).

Coffey M. On a Potntial with the Spectrum of the Hydrogen Atom. *J.Phys.A: Math. Gen.* 21, 3183 (1988).

Collas P. Algebraic Solution of the Kepler Problem Using the Runge-Lenz Vector. *Am.J.Phys.* 38, 253 (1970).

Collinson C.D. The Mathematical Simplicity of the Inverse Square Force. *Math.Gaz.* 61, 139 (1977).

Colwell P. Bessel Functions and Kepler's Equation. *Am. Math. Mon.* 99, 45 (1992).

Comberscure M. et al. Are N=1 and N=2 Supersymmetric Quantum Mechanics Equivalent? *J.Phys.A: Math.Gen.* 37, 10383 (2004).

Cooper F. et al. Supersymmetry and Quantum Mechanics. *Phys. Reports.*251, 271 (1995).

Cordani B. *On the Fock Quantization of the Hydrogen Atom.* Preprint 47/1988. Milano, Italia.

Correa F. and Plyushchay M. *Hidden Supersymmetry in Quantum bosonic Systems.* arXiv: hep-th/0605104.

Cotaescu I.I. and Visinescu M. *Infinite Loop Superalgebras of the Dirac Theory on the Euclidean Taub-NUT space.* arXiv: 0705.0866[hep-th].

Cotaescu I.I. et al. Runge-Lenz Operator for Dirac Field in Taub-NUT Background. *Phys.Lett.*B502,229 (2001).

Cotaescu I.I. et al., *Hierarchy of Dirac, Pauli and Klein-Gordon conserved operators in Taub-NUT background.* arXiv: hep-th/0107205.

Coulson C.A. and Joseph A. Self-Adjoint Lader Operators. *Rev. Mod. Phys.* 39,838 (1967).

Critchfield Ch. Scalar Potential in the Dirac Equation. *J.Math.Phys.* 17, 261 (1978).

D'Hoker E. Vinet L. *Spectrum (Super)-Symmetries of Particles in a Coulomb Potential.* CU-TP-292, 1985.

Dahl J.P. Physical Origin of the Runge-Lenz Vector. *J.Phys.A: Math. Phys.* 30, 831 (1997).

De Groot E.H. The Virial Theorem and the Dirac H Atom. *Am.J.Phys.* 50, 1141 (1982).

Dirac P.A.M. Forms of Relativistic Dynamics. Rev. *Mod. Phys.* 21,392 (1949).

Djemai A.E.F. *On Noncommutative Classical Mechanics.* arXiv: hep-th/0309034.

F.De Jonghe et al. *New Supersymmetry of the Monopole.* arXiv: hep-th/9507046.

Faux M. and Spector D. A BPS Interpretation of Share Invariance. *J.Phys.A: Math.Gen.* 37, 10397 (2004).

Fock V.A. *The Theory of Space, Time and Gravitation.* (Oxford, Pergamon), 1949.

Gel'fand I.M.and Graev M.I. *Irreducible Representations of Lie Algebra of U(p,q) Group*. In Proceedings "Physics of High Energy and Elementary Particles", N*aukova Dumka,* Kiev, 1967.p. 216.

Gendenstein L.E. and Krive I.V. Supersymmetry in Quantum Mechanics. *Sov. Phys*. Usp., 28, 645 (1985).

Giachetti R. and Sorache E. *The Discrete Spectrum of the Dirac Equation in Confining Potentials*. arXiv 0706.0127 [hep-th].

Goldstein H. et al. *Classical Mechanics,* 3rd Ed. Pearson Educ. Inc, 2004.

Goldstein H. More on the Prehistory of the Laplace or Runge-Lenz Vector. *Am.J.Phys*. 44, 1123 (1976).

Goldstein H. Prehistory of the "Runge-Lenz" Vector. *Am. J. Phys*., 43, 737 (1975).

Goodman B. and Ignjatovic. A Simpler Solution of the Dirac Equation in a Coulomb Potential. *Am.J.Phys*. 65, 214 (1997).

Gorringe V.M. *Generalizations of the Laplace-Runge-Lenz Vector in Classical Mechanics.Thesis*. Univ. Witwatersrand, Johannesburg, 1986.

Gorringe V.P. and Leach P.G.L. *Conserved Vectors and Orbit Equations for Motions in Two and Three Dimensions, in Finite-Dimensional Integrable Nonlinear Dynamical Systems,* World Scientific, Singapore, 1988.

Greenbrg D. *Accidental Degeneracy*.

Greiner W. Muller B. and Rafelski J. *Quantum Electrodynamics of Strong Fields*. Springer-Verlag. 1985.

Greiner W.and Muller B. *Quantum Mechanics. Symmetries*. Springer-Verlag, 1994.

Gu X.-Y. et al. Exact Solutions to the Dirac Equation for a Coulomb Potential in D+1 Dimensions. *Int.J.Mod.Phys*. E11, 335 (2002).

Habara Y. et al. *SupersymmetricRelativistic Quantum Mechanics*. arXiv: hep-th/0505237.

Hanc J. et al. *Symmetries and Conservation Laws: Consequences of Noether's*

Haymaker R.W. and Rau A.R.P. *Supersymmetry in Quantum Mechanics. Am.J.Phys*. 54, 928 (1986).

Heintz W.H. Runge-Lenz Vector for Nonrelativistic Kepler Motion Modified by an Inverse cube Force. *Am.J.Phys*. 44, 687 (1076).

Heinz W.H. Determination of the Runge-Lenz Vector. *Am.J.Phys*. 42,1078 (1974).

Hezel T.P. et al. Classical View of the Stark Effect in Hydrogen Atoms. *Am. J. Phys*. 60, 324 (1992).

Hietarinta J. Classical versus Quantum Integrability. *J. Mah. Phys*. 25, 1833(1984).

Hiller J.R. *Solution of the One-Dimensional Dirac Equation with a Linear Scalar Potential*. Am.J.Phys. 70, 522(2002).

Ho C.-L. *Quasi-exact Solvability of Dirac Equation with Lorentz Scalar Potential*. arXiv: hep-th/0511204.

Holas A. and March N.H. A Generalization of the Runge-Lenz Constant of Classical Motion in a Central Potential. *J.Phys.A: Math*.Gen. 23,735 (1990).

Horvathy P.A. et al. *Monopole Supersymmetries and the Biedenharn Operator*. arXiv: hep-th/0006118.

Horvathy P.A. *New Applications of the Biedenharn-Temple Operator*. arXiv: hep-th/0410161.

Horvathy P.A. *The Biedenharn Approach to Relativistic Coulomb-Type Prphlems*. arXiv: hep-th/0601123.

IUHET 227.

Iwai T. and Katayama N. Mltifold Kepler Systems – Dynamical Systems All of Whose Bounded Trajectories Are Closed. *J.Math.Phys.* 36, 1790 (1995).

Jackiw R. Dynamical Symmetry of the Magnetic Monopole. *Ann.Phys.* (NY).129, 183 (1980).

Jarvis P.D. and Stedman G.E. Supersymmetry in Second-Order Relativistic Equations for the Hydrogen Atom. *J.Phys. A: Math. Gen.* 19, 1373 (1986).

Jezewski D.J. and Mittleman D. Integrals of Motion for the Classical Two-Body Problem with Drag. *Int. J. Nonlin. Math.* 18, 119 (1983).

Kaplan H. The Runge-Lenz Vector as an "extra" Constant of the Motion. *Am. J. Phys.* 54, 157 (1986).

Katsura H. and Aoki H. Exact Supersymmetry in the Relativistic Hydrogen Atom in Genral Dimensions – Supercharge and the Generalized Johnson-Lippmann Operator. *J.Math.Phys.* 47, 032302 (2006).

Katzin G.H. and Levine J. Time Dependent Quadratic Constant of Motion, Symmetries, and Orbit Equations for Classical Particle Dynamical Systems with Time-Dependent Kepler Potantials. *J.Math.Phys.* 24, 552 (1982).

Khachidze T.T. and Khelashvili A.A. An "Accidental" Symmetry Operator for the Dirac Equation in the Coulomb Potential. *Mod.Phys.Lett.* 20, 2277 (2005).

Khachidze T.T. and Khelashvili A.A.. Hidden Symmetry Operator of the Kepler Problem in Relativistic Quantum Mechanics – from Pauli to Dirac, *Am.J.Phys.* 74, 628 (2006).

Khachidze T.T. and Nadareishvili T.P. Dirac Equation and its Squared Form. *Bull. Georg. Acad. Sci.*, 174, 61 (2006).

Khachidze T.T.and Khelashvili A.A. Alebraic Derivation of the Spectrum of the Dirac Hamiltonian for an arbitrary combination of the Lorentz-Scalar and Lorentz-Vector Coulomb Potentials. *Ukr. Fiz. Journ.* 52, 421 2007).

Khachidze T.T.and Khelashvili A.A., *Supercharge Operator for Hidden Symmetry in the Dirac Equation.,* In Proceedings of CICHEP II. Vol. 881, 279. (Ed. S.Khalil), Melville, New York, 2007.

Khare A. *Supersymmetry in Quantum Mechanics.* arXiv: math-ph/0409003.

Kirchberg A. et al. *Algebraic Solution of the Supersymmetric Hydrogen Atom in d Dimensions.* arXiv: hep-th/0208228.

Kirchberg A. et al. *Extended Supersymmetries and the Dirac Operator.* FSU-TPI 01/04. arXiv: hep-th/0401134.

Kirchberg et al. Algebrai Solution of the Supersymmetric Hudrogen Atom in d dimensions. *Ann. Phys.*(NY). 303, 359 (2003).

Kostelesky V.A. *Atomic Supersymmetry, Oscillators, and the Penning Trap.*

Kostelesky V.A. et al. *Supersymmetry and the Relationship between the Coulomb and Oscillator Problems in Arbitrary Dimensions.* Preprint LA-UR- 84-3972.

Kowen M. and Mathur H. On Feynman's Analysis of the Geometry of Keplerian orbits. Am.J.Phys. 71, 397 (2003).

Krause J. On the Complete Symmetry Group of Classical Kepler System. *J. Math. Phys.* 35, 5734 (1994).

Leach P.G.L. and Flessas. Generalizations of the Lapace-Runge-Lenz Vector. *Journ. Nonlin. Math. Phys.* 10. 340 (2003).

Leach P.G.L. Application of the Lie Theory of Extended Groups in Hamiltonian Mechanics for Oscillator and the Kepler Problem. *J.Austral. Math. Soc.* B23, 173 (1981).

Leach P.G.L. The First Integrals andOrbit Equation for the Kepler Problem with Drag. *J.Phys.A: Math.Gen.* 20, 1997 (1987)

Levai G. Solvable Potentials Associated with su(1.1) Algebras: a Systematic Study. *J.Phys.A: Math. Gen.* 27,3809 (!994).

Leviatan A. Pseudospin, Spin, and Coulomb-Dirac Symmetries: Doublet Structure and Supersymmetric Patterns. *Int. J. Mod. phys.* E14: 111 (2005).

Leviatan A. *Supersymmetric Patterns in the Pseudospin Spin and Coulomb Limits of the Dirac Equation with Scalar and Vector Potentials.* arXiv: nucl-th/0312018.

Lewis H. and Ralph Jr. Class of Exact Invariants for Classical and Quantum Time-Dependent Harmonic Oscillators. *J.Math.Phys.* 9, 1976 (1968).

Lewis H. and Ralph Jr. Classical and Quantum Systems with Time-Dependent Harmonic Oscillator Type Hamiltonians. *Phys. Rev. Lett.* 18, 510 (1967).

Lutzky M. Dynamical Symmtries and Conserved uantities. *J.Phys.A: Math.Gen.* 12, 973 (1979).

Martinez-y-Romero R.P. et al. *Relativistic Quantum Mechanics of a Dirac Oscillator.* arXiv: quant-ph/9908069.

Martinez-y-Romero R.P. Relativistic Hydogen Atom Revisited. *Am.J.Phys.* 68, 1050 (2000).

Mavodyan L et al. *Relationship Between Quantum Mechainics with and without Monopoles.* arXiv: hep-th/0610301.

McCauley J.L., *Classical Mechanics,* Cambridge, 1998.

McIntosh .V. and Cisneros A. Degeneracy in the Presence of a Magnetic Monopole. *J.Math.Phys.* 11,896 (1970).

McIntosh H.V. "*Symmetry and Degeneracy"*, In "Group Theory and its Applications" (ed. E.M.Loebl), Vol.II, pp.75-144.Acad.Press, New York, 1971.

Meng G. *Generalized Kepler Problms.* arXiv: math-ph/0509002.

Meng G. *The MICZ-Kepler Problems in All Dimensions.* arXiv: math-ph/0507028.

Menon V.J. Spin Conservation in One-Particle Dirac Theory. *Am.J.Phys.* 60, 928 (1992).

Mitchell T.P. Vector Constants and Their Algebras for Classical Hamiltonians. *J.Math.Phys.* 19, 2082 (1978).

Mladenov I.M. and Tsanov V.V. Geometric Quantizaion of the MIC-Kepler Problem. J.Phys.A: *Math. Gen.* 20, 5865 (1987).

Montgomery W. and O'Raifeartaigh L. Noncompact Lie-algebraic Approach to the Unitary Representations of SU(1.1): Role of the Confluent Hypergeometric Equation. *J.Math.Phys.* 15,380 (1974).

Moreira I.C. et al. Lie Symmetries for the Charge-Monopole Problem. *J.Phys.A: Math.* Gen. 18,L427 (1985).

Moshinsky M. and Szczepaniak. Th Dirac Oscillator. J.Phys.A: Math. Gen. 22, L817 (1989).

Motz L. The Conservation Principles and Kepler's Laws of Planetary Motion. *Am.J.Phys.* 43, 575 (1975).

Munoz G. *A Second Constant of the Motion for Twq-Dimensional Positronium in a Magnetic Field.* arXiv: physics/0303105.

Naymark M.A. *Unitary Representations of non-Compact Gpoups.* In Proc. "Physics of High Energy and Elementary Particles", *Naukova Dumka*, Kiev,1967, p. 183.

Nersessian A. *Generalization of MICZ-Kepler System.* arXiv: 0711.1037[math-ph], 2007.

Ng Y.J. and Dam van H. *A Geometrical Derivation of th Dirac Equation.* arXiv: hep-th/0211002

Novozhilov V. and Yuri. *Klein-Fock Equation, Proper-Time Formalism and Symmetry of Hydrogen Atom.* hep-ph/9911378.

Patera R.P. Momentum-Space Derivation of the Runge-Lenz Vector. *J.Phys.*49, 593 (1981).

Patera R.P. Scattering Supplement to "Velocity Space and Geometry of Planetary Orbits". *Am.J.Phys.* 46, 854 (1978).

Plyushchay M. Nonlinear Supersymmetry: from Classical to Quantum Mechanics. *J.Phys.A: Math.Gen.* 37, 10375 (2004).

Plyushchay M.S. *Monopole Chern-Simons Term: Charge-Monopole System as a Particle with Spin.* arXiv: hep-th/0004032.

Plyushchay M.S. On the Nature of Fermion-Monopole Supersymmetry. *Phys.Lett.* B485:187(2000.

Popov V.S., On Hidden Siymmetry of the Hydrogen Atom", In *Proceedings of Int. School "Physics of High Energy and Elementary Particles"*, Naukova Dumka, Kiev, 1967, p.702.

Rau A.R.P. Supersymmetry in Quantum Mechanics: An Extended View. *J.Phys. A: Math.Gen.* 37, 10421 (2004).

Romero J.M. and Vergara J.D. *The Kepler Problem and Non-Commutativity.* (2004).

Roy T. The Energy Eigenvalues of the Dirac Hydrogen Atom. *Am. J. Phys.* 56, 379 (1988).

Semenov V.V. Supersymmetry and the Dirac Equation of a Neutral Particle with an Anomalous Magnetic Moment in a entral Electrostatic Field. *J.Phys.A*:*Math. Gen.* 23, L721 (1990).

Serebrennikov V.B. Shabad A.E., Method of Calculation of the Radial- Symmetric Hamiltonian Spectrum on the Basis of Approximated O(4) and SU(3)-Symmetry, *Theor. Math.Phys.*,8,644 (1971).

Speise D. The Kepler Problem from Newton to Johan Bernulli. *Arch. Hist. Sci.* 50, 103 (1996).

Stahlhofen A.A., Algebraic Solutions of Relativistic Coulomb Problems, Helv. *Phys. Acta*, 70, 372 (1997).

Stanciu G.N. Further Exact Solutions of the Dirac Equation. *J.Math.Phys*. 8, 2043 (1967).

Stickforth J. The Classical Kepler Problem in omentum Space. *Am.J.Phys*. 46, 74 (1978).

Su J.-Y. Simplified Solutions of the Dirac-Coulomb Equation. *Phys. Rev.* A32, 3251 (1985).

Sukumar C.V. Supersymmetric Quantum Mechanics and the Inverse Scattering Method. *J.Phys. A: Math. Gen.* 18,2937 (1985).

Sukumar C.V. Supersymmetric Quantum Mechanics of One-Dimensional Systems. *J.Phys. A: math. Gen.* 18, 2917 (1985).

Tangerman R.D. and Tjon J. *A., Exact Supersymmetry in the Nonrelativistic Hydrogen Atom, Phys. Rev.* A 48, 1089 (1993).

Tempesta P. et al. *Exact Solvability of Superinegrable Systems.* preprint LPT-ORSAY 00-90. hep-th/0011209.

Thaller B. *The Dirac Equation.* Springer-Verlag. 1992.

Theorem. *Am.J.Phys.* 77, 428 (2004).

Thompson G. Polynomial Constants of Motion in Flat Space. *J.Math.Phys.* 25, 3474 (1984).

Tikochinsky Y. A Simplified Proof of Bertrand's Theorem, *Am.J.Phys.* 56, 1073 (1988).

Tompson G. Second Order Systems with Runge-Lenz-Type Vectors. *Lett.Math.Phys.* 14, 69 (1987).

Turek M. et al. Spin-Orbit Pendulum (Relativistic Extension). *Acta Phys. Pol.* B31, 517 (2000).

Valent G. The Hydrogen Atom in Electric and Magnetic Fields: Pauli's 1926 Article. *Am.J.Phys.* 71, 171 (2001).

Waldenstrom S. On the Dirac Equation for the Hydrogen Atom. *Am.J.Phys.* 47, 1098 (1979).

Weinberg S. *Quantum Theory of Fields.III. Supersymmetry.* Cambridge Univ. Press 2000.

Wipf A. *Algebraic Solution of the Supersymmetric Hydrogen Atom.* Quantum Theory and Symmetries IV. 1, Heron Press, Sofia, 2006.

Witten E. Spontaneous Symmetry Breaking in SUSY Quantum Mechanics. *Nucl.Phys.* B188, 513 (1981).

Wong M.K.F. and Yeh H.-Y. Simplified Solution of the Dirac Euation with a Coulomb Potential. *Phys.Rev.*D23, 3396 (1982).

Wong M.K.F. Infeld-Hull factorization and the Simplified Solution of the Dirac-Coulomb Equation. *Phys. Rev.* A34, 1559 (1986).

Wu J. et all. A Group Theoretical Calculation of the S-matrix for the Dirac-Coulomb Problem. *J.Phys.A: Math. Gen.* 20, 4637 (1987).

Yahiaoni S.-A. and Bentaiba M. *Exactly Solvable Potentials by so(2.2): Dynamical Algebra.* arXiv: 0711. 2265 [math-ph].

Yoshida T. A Vectorial Derivation of Kepler's Equation. *Am.J.Phys.* 56, 561 91988).

Yoshida T. Considerations on the Precessing Orbit via a Rotating Laplace- Runge-Lenz Vector. *Am.J.Phys.* 55, 1133 (1987).

Zachos C. and Curtright T. *Branes, Quantum Nambu Brackets, and the Hydrogen*

Zarmi Y. The Bertrand Theorem Revisited, *Am.J.Phys.* 70, 446 (2002).

J.Fris et al. On Higher Symmetries in Quantum Mechanics. *Phys.Lett.* 16, 354 (1965).

INDEX

A

B

C

S

T

U

V

W

Y